About Island Press

Since 1984, the nonprofit organization Island Press has been stimulating, shaping, and communicating ideas that are essential for solving environmental problems worldwide. With more than 1,000 titles in print and some 30 new releases each year, we are the nation's leading publisher on environmental issues. We identify innovative thinkers and emerging trends in the environmental field. We work with world-renowned experts and authors to develop cross-disciplinary solutions to environmental challenges.

Island Press designs and executes educational campaigns, in conjunction with our authors, to communicate their critical messages in print, in person, and online using the latest technologies, innovative programs, and the media. Our goal is to reach targeted audiences—scientists, policy makers, environmental advocates, urban planners, the media, and concerned citizens—with information that can be used to create the framework for long-term ecological health and human well-being.

Island Press gratefully acknowledges major support from The Bobolink Foundation, Caldera Foundation, The Curtis and Edith Munson Foundation, The Forrest C. and Frances H. Lattner Foundation, The JPB Foundation, The Kresge Foundation, The Summit Charitable Foundation, Inc., and many other generous organizations and individuals.

The opinions expressed in this book are those of the author(s) and do not necessarily reflect the views of our supporters.

HOLISTIC MANAGEMENT HANDBOOK

HOLISTIC MANAGEMENT HANDBOOK

Regenerating Your Land and Growing Your Profits

THIRD EDITION

Jody Butterfield,
Sam Bingham,
and Allan Savory

ISLANDPRESS   Washington | Covelo | London

The authors are grateful for permission to reprint the excerpt quoted in part 3 from Don Coyote: The Good Times and the Bad Times of a Much Maligned American Original, Dayton O. Hyde, Johnson books, 2004.

Holistic Management is a registered trademark of Holistic Management International. Island Press is a trademark of The Center for Resource Economics.

Library of Congress Control Number: 2018961802

All Island Press books are printed on environmentally responsible materials.

Manufactured in the United States of America
10 9 8 7 6

Keywords: Cash flow, context checks, drought reserve, ecological health indicators, growing season aide memoire, holistic context, Holistic Ecological Monitoring, Holistic Financial Planning, Holistic Land Planning, Holistic Management, Holistic Planned Grazing, infrastructure planning, land restoration, nongrowing season aide memoire, planning maps, profit, regenerative grazing, wildlife habitat.

CONTENTS

Part 3. Holistic Ecological Monitoring

Part 4. Holistic Land Planning

PREFACE

This book is designed as a practical companion for people who actually manage land and any others who want to know what responsibility for land really means. Most of all, however, it is written for the ranchers and farmers who make a living directly from the earth itself. Upon them rests the fate of everyone else.

The companion book to this one, *Holistic Management: A Commonsense Revolution to Restore Our Environment,* by Allan Savory with Jody Butterfield, attempts to clear the view from theory toward practice. This handbook looks the other way—from practice toward theory. In fact, however, both the theory and the practice evolved together and are still evolving, and you, too, should cut away at the thicket in your mind from both sides. The two books cover much of the same ground but from different perspectives. You need both. Don't risk resources on the information in this book without reading the textbook, and don't manage your operation from the advice in the textbook without mastering the details explained here.

How This Book Works

Allan Savory has long warned managers to "never think of Holistic Management as a management *system!*"

Easier said than done.

Although each of the four planning and monitoring procedures covered in this book takes a systematic approach to the management of your land, livestock, and finances, built into them is the realization that no land, no family, no economy, no marketplace, and no weather is the same year after year. The word *plan* becomes a twenty-four-letter word: *plan–monitor–control–replan.*

And that's what can save you as you strike the inevitable problems—a hiccup in the futures market, a prize bull that tears up fences, an in-law that destroys pickup trucks, drought. In the beginning, however, too many managers scrap the planning, monitoring, controlling, and replanning when things go wrong. It's easier to beg off and say, "Oh well, at least we've got more herd effect than before!"

No. The last of the old way always beats the first of the new. The first musket couldn't match the best crossbow. The first automobile trailed a good horse. Your first plans may be so far from reality that going back to seat-of-the-pants management appears to make a lot more sense than replanning. Sam Bingham tells a story that illustrates the point:

> I remember a woman in a Navajo adult education class who would not learn to use a ruler. When asked to draw a line through the center of a page, she produced a slash across one corner. "Oh, that's the way it came out with the ruler," she said. "Well, what about without the

ruler?" I asked, knowing that as an expert weaver her native gift for geometry far outstripped mine.

She squinted a second at the paper, then drew a line freehand that missed the mark by a millimeter top and bottom. Damn good, but I could still top that with a ruler, despite my inborn lack of talent. So, too, with the planning procedures of Holistic Management. Learn them, and you will soon beat raw intuition every time.

Holistic financial planning, grazing planning, ecological monitoring, and land planning cannot begin, however, without a good grip on some fundamental concepts and a clear idea of your specific priorities. Only after this groundwork can you begin the step-by-step process of generating a plan or, in the case of monitoring, documenting environmental change.

This book is therefore divided into four major parts, each with two sections. The first section in each part covers basic concepts and technique. The second section is the step-by-step procedure for getting your plan down on paper, monitoring it, and adjusting it; or, in the case of ecological monitoring, recording changes on the land and interpreting the results. In part 4, "Holistic Land Planning," the fundamental structure of basics followed by steps breaks down a bit because land planning is not bound to the strict annual cycle that governs finances, livestock, forage, and crops. Nevertheless, the ultimate objective—getting a comprehensive plan on paper—holds.

One final note. The figures used in the examples included in this book will be realistic to some but not to others because prices and costs vary so much the world over. Some figures do come from actual examples; others are merely imagined. Never assume that any figures can be translated to your own operation. You will always need to work them out for yourself.

Don't use this book until you have familiarized yourself thoroughly with the theory and science of Holistic Management. Read *Holistic Management* first, and if possible attend a training course, make contact with someone who has, or seek assistance from a Savory Network Hub near you.

Acknowledgments

The bulk of the ideas in this book came through Allan Savory, out of his long study of the land, and the work of many others besides. Feeling that he had grown stale in presenting material he had worked with for over thirty years at the time, he asked Sam Bingham to author the first edition of the handbook, which appeared in 1990. Sam had a flair for making dry material come to life and understood the struggles of those new to practicing Holistic Management, and helped inspire a whole generation of readers in that first edition. When he was unable to work on a second edition, Jody Butterfield stepped in to provide more than a decade's worth of updates. She was listed first on the title page because the updates were substantial, and that remains true for this edition. But because so much of the book is still in Sam's words and includes so many of his examples, his name also remains on the title page.

Most of the ideas and refinements in this third edition of the *Holistic Management Handbook* appeared first in the Savory Institute eBooks series, which provides annual updates to all of the planning and monitoring procedures based on the feedback and insights of practitioners and the Savory Accredited Professionals serving them. A special thanks to Accredited Professionals Sheila Cooke (UK) and Ulf Ullring (Norway) who provided additional material on grazing and land planning in nonbrittle environments, and to Savory Institute's Senior Program Director, Byron Shelton, who added his own insights and patiently reviewed every paragraph we sent him. Our deepest thanks to all of the ranchers and farmers who, in putting these ideas into practice, have shown what is possible and inspired us all.

PART 1

HOLISTIC FINANCIAL PLANNING

HOLISTIC FINANCIAL PLANNING
Generating Lasting Wealth

THIS BOOK DOES NOT BEGIN WITH LAND PLANNING, stock density, paddocks, and grazing periods. It begins with financial planning, because money is the ruler. For better or worse it is the ruler even though your holistic context goes far beyond profit. In fact, both the materialist who finds the breath of life in crisp new bills and the poet who would rather live without them might find more peace of mind and freedom of spirit by accepting the notion that money is nothing but a tool. It is certainly not fulfillment and not necessarily even wealth, broadly speaking.

Holistic thinking would be infinitely harder without the benefit of a tool like money by which to measure progress, though of course money is certainly not the only measurement. By definition holism deals with "wholes" in which many elements affect each other simultaneously, but the human brain can't handle everything at once. Since your wallet probably can't either, Holistic Financial Planning is the process through which you will reduce the grand notion of holism to the practical matter of what you do first and how much of it you do, so the whole will come out right.

That said, Holistic Financial Planning does not result in a business plan, but ideally should operate within one that includes multiple-year goals and strategies, performance milestones, and so on. Holistic Financial Planning will result in an annual statement of your income and expenses, which includes a budget and cash flow that you plan forward with an understanding of holistic decision making and key environmental insights. It will be accompanied by a statement of your beginning and ending net worth (similar to, or the same as, a balance sheet). It is not linked to an accounting system, nor monitored and controlled by your accountant. You are in charge of managing and monitoring the plan you create. However, the more you can match the account names used in your accounting system to the income and expense categories in your financial plan, the easier it will be to operate and manage the financial plan.

Two key principles in Holistic Financial Planning differentiate it from other methods:

- *Plan profit before planning expenses.* Just as work expands to fill the time available, so expenses often rise to the level of anticipated income. So, when creating the plan, we plan the profit first to overcome this tendency.
- *Check for context alignment.* Which actions are actually moving you closer to your goals? The context checks make sure you prioritize those actions that do move you in that direction while ensuring you achieve a healthy triple bottom line (financial, environmental, social).

The planning process includes two parts. The first is devoted to reviewing the current year's plan, gathering information and figures for the new plan, and running decisions through the context checks. All the thinking, decision checking, and sorting of the information you compile then comes into play in the second part when you put your plan on a spreadsheet. If you've done the first part well and researched the costs involved in the ideas you want to implement, you probably won't need more than a day or two to put your plan on paper.

Mastering the Basics

HOLISTIC FINANCIAL PLANNING is the single most important activity you can undertake each year to ensure that all the money you earn and spend is in line with your holistic context. If you seek prosperity and financial security, few activities during the year count more than this planning. It takes precedence over vacations, interruptions, and excuses of any kind.

Your Holistic Context: Aligning Your Financial Decisions

Your holistic context ties what you value most in life to your life support system. It starts with an expression of how you want your life to be in the whole you are managing and then describes the environment and behaviors that will sustain that quality of life for future generations. Keep it in mind at all times as you proceed through the Holistic Financial Planning process. Ambiguity about your holistic context may not result in an unworkable plan, but it could very well generate a plan you will not want to work (and may even sabotage subconsciously).

As money is merely a measurement, so profit is a means of fulfilling the desires expressed in your quality of life statement. A close family, the creation or preservation of good land, public service, church work, the education of your children, loyalty to relatives, and many other desires and duties all put demands on profit. If you do not have these things in mind when you plan your commitment of money and labor, you will make a plan that you will inevitably scrap the minute these other aspects of your holistic context demand it. Clarity in your holistic context will enable you to avoid temptations and opportunities of tremendous promise that nevertheless lead in the wrong direction for you.

The planning procedure, covered later on in "Creating Your Plan," will help you organize a huge amount of complex information about operations that go on simultaneously, but you still have to go one step at a time and put one thing ahead of another. Clarity in your holistic context and a deep ownership in it make that possible.

The Context Checks: Reducing Decision-Making Stress

At this point, reread *Holistic Management: A Commonsense Revolution to Restore Our Environment*, third edition, chapters 24 through 31 on the context checks: Cause and Effect, Weak Link, Marginal Reaction, Gross Profit Analysis, Energy/Money Source and Use, Sustainability, and Gut Feel. A brief summary also appears at the end of this book in appendix 1. In Holistic Financial Planning, all your policies and projects must come up for review through these checks as you allocate resources. You may start with a hundred different ideas, but before you actually plan action on any of them, you have to evaluate their soundness and set priorities. What enterprises, what investments in land improvements, what training for your staff, and so forth, will you try to carry out this year with the resources you have? Use the context checks when you

SIDEBAR 1-1

A Word about Profitability

In conventional planning, managers first plan production—crops, meat, timber, hunting leases, and so on. Then they calculate the anticipated income for the year, then the expenses, and finally cut and paste until they see a positive balance. This tends to make profitability the ultimate test, for which all other considerations are compromised.

Holistic Financial Planning proceeds in a rather different way. Profit is planned before any expenses, and it ranks alongside other elements that will sustain your definition of a good quality of life, which probably includes prosperity or economic security. The context checks include a "gross profit analysis," which is designed to highlight how much each enterprise contributes to covering fixed costs, or overhead. This ensures that a plan will indeed produce profit. The other checks then help ensure that the actions you take to create that profit are socially, environmentally, and economically sound—simultaneously—both short and long term.

to find the underlying cause of a problem. If the potential new enterprise passes the other checks, a more detailed gross profit analysis will be needed.

As neat as the context checks appear in theory, they overlap a good deal, and for good reason: what you might miss in one you pick up in another. Sometimes it proves impossible to figure out where one or another applies, but of the seven checks, a few will almost always prove critical in a given case. Some, such as sustainability, almost always apply. Don't agonize over ambiguity in regard to any one check. They all function together like the elements in one of those filters that purifies water through a series of screens, flotations, and catalysts, each of which eliminates one class of contaminants while ignoring the rest.

It is important to remember why you are doing what you are doing, so let's recap. You are attempting to make decisions that are economically, socially, and environmentally sound and aligned with your holistic context. The checking questions help you do that. They come into play in Holistic Financial Planning because in deciding where to allocate money, you are actually making most of the major decisions for the year.

Holistic Management covers cause and effect, sustainability, and gut feel well enough to warrant no further practical advice here. The same goes for weak links in the social and biological contexts. The other checks, however, require some translation into dollars and cents. Let's take them one by one.

Financial Weak Link: Generating Wealth

The financial weak link check has profound implications for deciding the fundamental question of planning: "How do I maximize the income I can generate?" The place to focus is on the weakest link in each enterprise. At any given moment there is only one weakest link, and you must deal with it before considering any other link.

You are trying to build a business and an environment (future resource base) that will endure, so you and following generations can sustain a profit. To do that, you want to ensure every year that your major investments of money and labor keep strengthening the weak link in the chain of production that exists at any point in time for each of the enterprises you engage in. Once you have

do this. They will cut a significant amount of the confusion out of this task.

Remember, when checking any action or decision to be clear about what it is you are checking. Don't rush to check a decision before you have clarified what your objective is—to reduce weeds in a pasture, buy a tractor, or whatever—and discussed all the aspects you normally would: what you know or need to know about the proposed action in terms of past experience, research results, a friend's advice, expert opinion, or what it will cost. Only then should you run through the context checks to make sure that in achieving your objective you remain aligned with your holistic context.

Speed is essential to the checking process, or you risk losing sight of the whole. However, two checks, cause and effect and gross profit analysis, require a lot of thought and, in the latter case, calculations using pencil and paper. Do them first if you are dealing with a problem (cause and effect) or assessing a possible new enterprise (gross profit analysis). Then pass quickly through all the other checks that apply. You may need to return to both these checks. The cause and effect check may require additional probing identified the year's weak link in each enterprise, you look at all the actions you could take that would strengthen that particular weak link as soon as possible. When allocating money for expenses, those actions that address the weak link in an enterprise will receive priority, assuming that other checks have been passed and the action is in line with your holistic context. Expenditures that address a weak link are considered wealth generating because they boost production, and thus profit, to a new level, though perhaps not until the following year. Other expenditures generally maintain production at current levels.

The chain of production always has three links, as shown in figure 1-1. If you are a rancher or farmer, your primary production is based on the conversion of sunlight energy (through plants) to a salable or consumable product, such as food, fiber, lumber, wildlife, ecosystem services, or recreation. Jargon aside, this means that plants capture solar energy to make food, then you turn that "food" into a marketable product—bale of hay; gallon of milk; bird-watching opportunities; a water, hunting, or fishing lease; or whatever. But you have to actually market that product or service before you have a dollar in your hand that you can live on or reinvest in the business. Ideally, you want the money you invest in a weak link to be in the form of solar dollars, which ranchers and farmers produce through harvesting sunlight and converting that

sunlight to money, as long as soils are not damaged in the process. Mineral dollars achieved at the expense of lost soil, or paper dollars borrowed from the bank can be used to strengthen a weak link, but in the first case they could undermine your future resource base, and in the second they usually come with interest attached.

Although the weak link in each enterprise shifts from year to year and can shift within the year, Holistic Management training programs for ranchers and farmers put great emphasis on the resource conversion link because ranchers and farmers have a tendency to leave that one to Nature, assuming management can do little to influence the amount of sunlight plants convert to usable energy.

But the rancher now knows that she can increase energy flow by using animal impact to break up soil capping and lay down litter to improve the water cycle and grow more plants, and plan the grazings to prevent overgrazing and enhance growth rates and growing time. The farmer can maximize energy conversion by good selection of heat- and cold-tolerant crops and planting dates, which can extend effective growing seasons, and by keeping soils covered, and creating good drainage, crumb structure, and abundant organic matter within them to maximize the area of leaf open and exposed to sunlight.

> The key to generating wealth is not in the things you sell, but in how you reinvest the money earned.

You can probably produce a landscape that will turn more sunlight into edible carbohydrates and protein, or farm in a way that builds soil fertility and reduces vulnerability to drought. Each enterprise you manage will always have a weak link at any moment in time. Thus, if you run cattle and sheep and plant sorghum, winter wheat, and Bermuda onions, you will have a weak link to find in five different enterprises. (See box 1-1 for more examples.)

Whatever you determine to be a weak link, you must set in motion an action plan to strengthen it. This action plan will show up on your financial planning sheets as a separate expense item for training, fencing, advertising, or whatever strengthens that link in the chain.

The Chain of Production

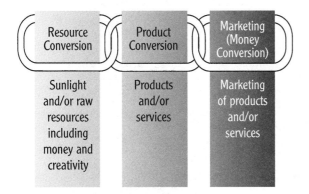

Figure 1-1. Human creativity first needs to utilize raw resources—sunlight in particular—and money to create a product or service. Then the product or service needs to be perfected and finally marketed to produce money. The chain is only as strong as its weakest link.

Box 1-1. Identifying the Weak Link

Here are some things to look for when trying to determine the weak link in the chain of production in a given year.

LIVESTOCK OPERATION

Resource (Sunlight Energy) Conversion
- Forage shortfall*
- Paddocks too few, or herd control too poor, to minimize overgrazing
- Too many herds
- Drainage poor
- Plant species composition poor
- Low litter accumulation
- High supplement cost
- Inadequate or unbalanced soil fertility

Product Conversion
- Unutilized forage that is oxidizing
- Sufficient forage but
  low calving/lambing rate
  poor gains
  poor genetics
  high mortality

Marketing (money conversion)
- Low prices
- Market resistance due to
  insensitivity to demand
  inadequate research
  poor quality
  poor sales effort
  ignorance of market mechanisms
  (futures, etc.)

CROP FARMING

Resource (Sunlight Energy) Conversion
- Acreage too small*
- Inputs too high (fertilizer, etc.)
- Poor water cycle management
- Drainage poor
- Overirrigation
- Planting only one crop when two or more crops could be grown in a field in the same year
- Planting too late
- Monoculture cropping
- Poor germination
- Poor crop health
- Inadequate or unbalanced soil fertility

Product Conversion
- High damage loss (insects, disease)
- Low-tonnage marketable product versus dry matter produced
- High harvest and handling loss
- Excessive crop-drying costs
- Transport damage

Marketing (money conversion)
- Low prices
- Market resistance due to
  inadequate research
  poor quality
  poor packaging
  poor sales effort
  ignorance of market mechanisms (futures, etc.)
  unnecessary middlemen

In a grazing operation, if forage production per acre is low, then increasing acreage doesn't help. "Forage shortfall" would be the weak link. On a crop farm, production per acre may be high, but acreage may be too small to generate sufficient revenue, and "acreage too small" would be the weak link.

Even though purchasing feed allows you to carry more animals, it does not strengthen the resource conversion link in terms of harvesting more sunlight; however, fencing, land acquisition, plantings, or improved drainage could. If you have too few animals to consume the forage you produce, then product conversion is the weak link, and a plan for increasing animal numbers addresses that link.

If the resource conversion link is weak for the hay enterprise, an investment in a new swather (windrower) or engine overhaul would be a waste of money. A new drainage system makes more sense because it enhances growth and thus solar energy conversion. Therefore, if your equipment will last another year, put the money into tile pipe instead.

There is often a gray area between the product and marketing conversion links. For example, poor-quality wool might be considered a product conversion weak link if it was tied to the quality of the animals that produced it, but a marketing conversion weak link if it was delivered dirty. What matters in these cases is not the precise positioning but that you see the problem and address it as you reinvest in the business.

You may have enterprises that are not directly dependent on solar energy conversion, as is the case with most urban businesses. A bed and breakfast enterprise on a corner of the property would be such an example. In this case the resource conversion link does not involve the conversion of sunlight. The resources you are converting to a product are money and creativity and might include building materials and the retired couple from town who will run the operation.

Energy/Money Source and Use: Investing Soundly

Money derived from the mineral wealth of the earth we term mineral (or petrochemical) dollars, and money derived through plants grown by the power of the sun we term solar dollars. A characteristic of mineral dollars is that we can choose either to use the resource from which they are produced over and over in a cyclical manner, or to mine and consume the resource all at once. Soil would be an example. We could choose to build soil and continue to use it for centuries, or we could destroy and erode it until it is gone. Tragically, mainstream agriculture has chosen the second option in most instances and has become an extractive industry, destroying more tons of soil than it produces in food each year.

Solar dollars are produced directly or indirectly from plants. As plants, like humans, are totally dependent upon soil, a farmer is not strictly producing solar dollars unless soil life, and thus soil, is maintained or enhanced.

The third form of money, paper dollars, is based upon human creativity and financial transactions, and its basis is no deeper or more solid than the public's confidence in the economy and the government. This form of money is very unstable and can be created or destroyed as land

values, interest rates, stock prices, inflation, and currency exchange rates fluctuate.

Ranchers or farmers are better off measuring their success in solar dollars only and relying on the paper ones at their peril. The argument is both moral and practical. The health and fortune of humankind cannot be sustained without the creation of solar dollars, and those engaged in the management of natural resources accept a special responsibility in this regard. On the practical side, the more you can rely on solar dollars, the more you insulate yourself from swings in land and commodity prices, interest rates, and the like. This does not mean ignoring such matters. In fact, it demands a particularly nimble and flexible attitude toward them.

All borrowing, especially if it is based on the current real estate market, involves paper dollars to some extent, so obviously your position is more stable if you can finance your plans out of solar dollars generated by your operation.

The second aspect of this check asks whether the planned use of energy or money is consumptive or cyclical. Is the use good only once, or is it durable in that it builds infrastructure for future wealth? As the wording of the check—energy/money source and use—indicates, you must apply the check to the source of energy and money you use and to the use of the energy and money that you generate, and you must consider both aspects in terms of your holistic context. For example, the use of diesel fuel in clearing land represents a consumptive use of a nonrenewable energy resource produced elsewhere. However, done once to produce a new cropland it could be in line with your holistic context, as it is building the infrastructure needed. The use of fossil fuel to aerially spray weeds infesting your land would constitute an addictive use of that resource unless you had first removed the cause of the weed infestation. Clear-cutting timber to make a loan payment is a consumptive use of solar wealth if it destroys the productivity of the soil. Better forestry practices or conversion into sustainable pasture might pass this check.

It is particularly helpful to apply this context check as you weigh ideas for strengthening the weak link in the chain of production. Of dozens of ways to proceed, some will represent a better use of energy and money than others. At the product conversion level, many rationalize heavy grain feeding to push livestock to a higher market class. Though the cash return may appear too good to refuse, the feed typically represents a consumptive use of soil and fossil energy. Finding a different market might reduce the need or even eliminate it altogether.

Marginal Reaction: Getting the Biggest Bang for Your Buck

To some degree the marginal reaction check is the accountant's equivalent of the weak link check, but it applies in many nonquantifiable situations as well. If you followed the idea perfectly, you would make your investments one dollar or one hour at a time, asking for each, Which investment will move me the furthest toward my goal? Each dollar (or hour) goes where it will yield most, and this changes whenever diminishing returns on one investment drop below what the next dollar or hour might return somewhere else. The only pitfall in practice is an expectation that you can quantify every situation, which of course you can't. In such cases this check will always be a subjective one.

Assuming that two or more different actions would help strengthen a particular weak link, there is little or no possibility that both or all of those actions would provide the same return toward the goal for each additional dollar or human hour invested. Since you are addressing a weak link, and every dollar matters, you use this check to determine, to the best of your ability, which of the alternatives gets all the dollars you can assign to it this year. Most commonly, you will be looking at marginal reaction per dollar or human hour of effort. However, if some other resource was limited, you could at times compare marginal reaction per unit of that limited resource.

Suppose resource conversion is your weak link. You can buy land, lease land, develop water on land not now accessible, reseed old land, hire herders or build fence, or retrain your staff. Which will have the greatest marginal reaction?

Advertising may strengthen a weak marketing link if you have to establish a market. After a point, however, extra publicity has only a short-term effect. Then storage facilities to gain market flexibility or better breeding stock to improve quality might have a greater marginal reac-

tion. They might enable you to enter new markets and stop wearing yourself out holding your share of a limited one.

Perhaps none of these strategies generates an acceptable marginal reaction in relation to the quality of life expressed in your holistic context. More carrying capacity or better sales might increase your income, but your daughter needs a loan to buy the place next door. Maybe that will pay back more in joy than the same money ever could in beef. The matter is further complicated by the old truth that time is money (and often much more than money). You must consider the marginal reaction of time. Will the hour spent checking fence return more in terms of your holistic context than the hour spent planning? How about spending the same hour with your family?

Think marginal reaction. You are building a life and a landscape and sources of wealth that will sustain you as far down the road as you dare look. You have to figure how far down that road each dollar or hour will send you.

Gross Profit Analysis: Bringing in the Most Money for the Least Additional Cost

Simply stated, gross profit analysis is a technique that separates *fixed costs* (defined here as costs you have regardless of what or how much you produce) from the costs directly linked to production, termed *direct* or *variable costs*. It enables you to compare many enterprises or combinations of enterprises. Besides expenses normally referred to as overhead, fixed costs can include other expenses that have already been incurred. For example, fertilizer—normally thought of as a variable cost—can be fixed if already bought and you are making a new decision about crops.

This approach, developed by David Wallace, a British agricultural economist, respects the fact that at any point many costs, such as living expenses, debt payments, and full-time labor, exist no matter what you produce. So they should not be considered in judging the potential of an enterprise you undertake tomorrow.

Many people do not make this distinction when analyzing the sources of profit and loss. They look only at the bottom line—which tells them little or nothing about which elements of an operation contribute most to covering their fixed costs and ultimately whether those fixed costs are justified. Even when different enterprises in an operation are analyzed separately, it is common to apportion some fixed costs among them. But that obscures the real contribution of that enterprise toward covering the total fixed costs of the business. People not used to this concept often find the distinction between fixed and variable expenses confusing and want a list. Again, any expense that you cannot avoid in the planning period, regardless of what you plan, is a fixed expense. Any expense that derives from your plan is variable.

Failure to see the difference can lead to disastrous decisions. The two graphs shown in figure 1-2, for example, illustrate the large-scale ruin of dairy operations in the mountains of western North Carolina. Beginning in the 1960s, industry experts encouraged local dairymen to build continuous milking parlors in association with feedlots supplied by intensive crop production or purchased feed. In theory this would produce more profit per cow—and virtually no end to the number of cows you could have.

The left-hand graph in figure 1-2 shows how traditional accounting justified this strategy. Cost per cow is enormous, but so is production, so gross profit per ani-

mal looks good enough to justify the large capital investment in the new equipment and feed purchase or production. Everything "cash flows." The centuries-old practice of grazing cows on steep mountain pastures looks pretty poor in comparison.

At the time of the decision, however, all investments in the continuous milking scheme were variable costs. These included such obviously variable costs as feed, vet bills, and semen to produce high-production cows. The cost of the new facilities also counted and should have been calculated as part of the annual cost per cow over the life of the facilities.

The most significant fixed costs were mortgage payments on land, and taxes, which were high because booming second-home development had inflated values. Since only a fraction of the mountain land was suitable for crops and feedlots, little of this mortgage or tax cost was assigned to the intensive milking systems, even though it had to be paid. Meanwhile, huge areas of pasture, laboriously cleared in past decades, went out of production under the new system and in that nonbrittle environment quickly

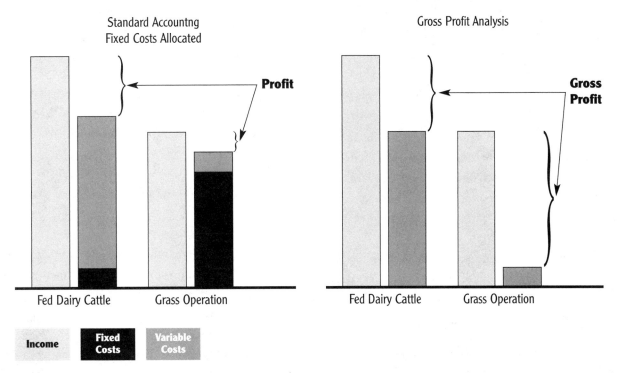

Gross Profit Analysis of Fed Versus Pastured Dairy Cattle

Figure 1-2. The left-hand columns compare profits when fixed land costs are allocated according to the amount of land used. The right-hand columns compare gross profit when land costs, which have to be paid anyway, are left out.

succeeded to forest. There was no going back. Those who could sell land to developers survived financially, but the dairy herds disappeared.

Gross profit analysis would have revealed the problem before the fatal step, as the second graph shows. If variable costs alone are considered, it becomes obvious that each cow on the old mountain grass went much farther toward covering the cost of the land and incurred relatively little "per head" expense compared to the cow on feed. Even when the high taxes ate up profit, die-hard farmers could often supplement their income enough to hold on to their property because the whole farm produced. Gross profit per acre remained healthy, and other inputs were low. The story might have turned out quite differently on flat, fertile land. But in that case a gross profit analysis might well have led to truck farming instead of intensive feed crops and Holsteins.

Gross profit analysis tends to influence your planning in three key ways:

1. You can be far more nimble in response to paper dollar changes. You will know when buying land makes sense, when sheep will do better than cattle, when yearlings beat calves, when to grow hay and when to buy it.

2. You will make much better use of the land, equipment, and labor you already have. The common practice masks the expense of idle assets. Gross profit analysis shows when you are better off making smarter use of what you have than making "cost-effective" new investments.

3. It makes you extremely aware of fixed costs. In the long run, none of those costs is really fixed. You can sell the ranch or farm itself if you have to. Short of that you can bias your planning to cut fixed costs and get rid of inefficient assets. Thus you can cut your exposure to the dangers of paper dollars—and compound interest—the great trap of agriculture in the United States and those countries following its lead.

Box 1-2 shows the difference this analysis makes to the priorities for profit in most livestock operations. The following simplified version of a real situation illustrates the technique for comparing enterprises using a gross profit analysis. The numbers are in South African rand, not dollars, and the relationship between income and animal numbers does not transfer, either, so do not expect the conclusions to relate to your operation. The method, however, will. In this particular case, two brothers differed as radically as Cain and Abel on the subject of sheep versus registered (or pedigree) cattle. The analysis answered their pressing economic question but not, alas, the human one.

The income and expense account shown in figure 1-3 has had a lot of detail boiled out of it, but it still includes many items that do not figure in the gross profit analysis. These represent items that either have nothing to do with the enterprises in question or are not "variable," in the sense that you will not get rid of them in the near future regardless of your conclusions about the various enterprises. Note that the analysis side has some categories, such as pasture, silage, and hay, that do not rate a line on the traditional account but are derived from several of the expense headings.

To compare the livestock options effectively, gross profit analysis results must be reduced to some common and comparable base—dollars per head per year, dollars per acre or hectare per year, dollars per standard animal unit per year, dollars returned per dollar invested per year, and so forth.

In this case, we are given the following:

1,500 sheep

400 commercial (grade) cattle

150 registered (pedigree) cattle

The gross profit per head (in rand) would be:

R511,780 ÷ 1,500 = R341.19/sheep

R222,470 ÷ 400 = R556.18/commercial cow

(R53,010) ÷ 150 = (R353.40)/registered cow (loss)

If the stocking rate reflects the commonly used ratio of five sheep to one cow unit, then the difference in gross profit per unit of land is drastic indeed. Sheep return R1,705.95 (R341.19 × 5) per animal unit in this particular case; registered cows lose R353.40.

Gross profit analysis helps greatly in determining which enterprises to engage in and also in monitoring the performance of each enterprise. However, all the actions

Box 1-2. Priorities for Profit in a Livestock Operation

If you are engaged in livestock production, it's important to understand the relationship between the number of animals you run on the land (stocking rate) and profitability. Due to the belief that overgrazing was a result of overstocking, ranchers have long looked to individual animal performance (conception rates, weight gains, etc.) for profit—to their detriment. Now that we address overgrazing by planning grazing time, the picture changes dramatically and so do the priorities for profit. For a cow-calf producer, for example, the following is now the actual order of priority for profit:

1. Stocking rate
2. Conception and calving rates
3. Weight gains

Higher conception/calving rates and weight gains cannot compare with simply increasing stocking rate, as the graph shows dramatically.

Generally, the fixed or overhead costs are the highest costs, and they exist whether you run one animal or a thousand. Fixed costs would include the money tied up in the value of the land, salaries, the buildings, vehicles, fencing, and other developments. Then come the variable costs, which increase as the number of animals increases. Variable costs would include labor, supplements and feed, veterinary and marketing costs, and so on. At **Point A** the rancher would have no livestock and no income from animals, but would have the ranch and its costs. He or she would then "fully stock" the ranch—based on past thinking that linked numbers to overgrazing—to **point B**, and both the income (dotted line) and variable costs would increase. The income from livestock would rise considerably, but the *profit* (income minus fixed + variable costs) is only moderate.

That's the best we can do under conventional livestock management. Because the belief was that reducing livestock numbers would improve the land and increase forage production, improving animal performance has become the highest priority. Vast sums have been spent on techniques and practices to improve animal performance. However, the money ranchers spend on better breeds, supplements and feed, and other practices, is considered a variable cost. As these variable costs have risen, profitability has generally kept falling because stocking rates have continued to fall. The land has continued to deteriorate as plants are still being overgrazed and soils and plants overrested.

When ranchers learn to improve their land's productivity by keeping their animals bunched and moving under Holistic Planned Grazing, the picture changes dramatically. More animals are needed, not fewer. When we increase the stocking rate to **point C**—a doubling of animal numbers to the benefit of the land—profit increases substantially. While the variable costs increase greatly with a doubling of animal numbers, *the fixed costs don't change*. Income keeps rising *and so does profit*. In practice, the effect can be even greater than shown in the graph because the variable costs can also drop proportionately as the quality of feed on regenerating soils improves.

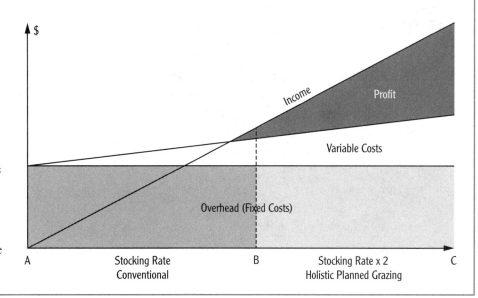

Gross Profit Analysis—Comparing Enterprises

Income & Expense Account		Gross Profit Analysis		
Income		**Sheep**	**Grade Beef**	**Registered Cattle**
Livestock operations				
Trading	1,277,760	696,900	283,620	297,240
Herd value change	(10,770)	(68,760)	207,000	(149,010)
Produce				
Wool	167,400	167,400		2,370*
Dairy	21,420	**R794,640**	**R490,620**	**R150,600**
Timber	2,160			
Miscellaneous				
Show prizes	2,370*			
Equipment sold	1,360			
Bonus and rebates	5,760			
Dividends	300			
Total Income	**R1,467,760**			

Gross Income from Enterprises Less Variable Costs

Expenses				
Feed, salt, etc.	241,260	65,880	86,340	89,040
Crops (seed and chemicals)	200,660	188,280 pasture		
Vehicle and equipment operation	193,200		72,000 silage	35,440
Repairs (fence and machines)	30,140		97,420 hay	48,000
Labor				
Seasonal	66,880			
Full-time	495,320			
Veterinary and A. I.	34,120	15,000	12,390	6,730
Shearing	14,600	14,600		
Registered stock fees	24,400			24,400
Electricity	15,900	R283,760	R268,150	R203,610
Insurance	41,160			
Administration	29,955			
Interest	81,390	R511,780	R222,470	(R53,010)
Miscellaneous	2,835			
Total Expense	**R1,472,020**			
Profit (Loss)	**(R5,060)**			

Gross Profit

* Show prizes relates to registered cattle.

Figure 1-3. End-of-year comparison of enterprises based on actual income and expenses. There are 1,500 sheep, 400 commercial cattle, and 150 registered cattle. In the shaded area, portions of the figures in the first column have been allocated to one or more enterprises because those amounts are considered a direct cost of that enterprise.

and technologies involved in an enterprise should also be checked to ensure they are in context. Use a gross profit analysis for three purposes:

1. To compare enterprises (as in the example)
2. To compare scenarios to see how different conditions, such as price changes, might affect an enterprise

3. To compare the use of assets to weigh the effect of diverting them from one enterprise to another

Comparing Scenarios
Gross profit analysis can be used to assess the impact of future events or policies. In the example shown in figure 1-4, "poor," "average," and "good" refer to price only. You

Gross Profit Analysis—Comparing Scenarios Based on Price

Gross Income	Poor	Average	Good
Weaning %	*85%*	*85%*	*85%*
Price/lb.	$ 1.25	$ 1.50	$ 1.75
Weaner wt/lbs.	*500 lbs.*	*500 lbs.*	*500 lbs.*
Weaner gross income	$ 53,125	$ 63,750	$ 74,375
Cull cows 20% (-2% mortality)	*18%*	*18%*	*18%*
Cull cow price	$ 500	$ 750	$ 1,000
Cull cow gross income	$ 9,000	$ 13,500	$ 18,000
Gross Income	**$62,125**	**$77,250**	**$92,375**
Direct Costs			
Replace heifers 20 @ $1,000, $1,500, and $2,000	$ 20,000	$ 30,000	$ 40,000
Veterinarian expense @ $10/(cow)	$ 1,000	$ 1,000	$ 1,000
Supplies (3 lbs./day for 3.5 months) (330 lbs. @ $160/ton)	$ 2,640	$ 2,640	$ 2,640
Interest @ 10% on average value $1,500	$ 15,000	$ 15,000	$ 15,000
Bull cost $12/(cow)	$ 1,200	$ 1,200	$ 1,200
Shipping	$ 0	$ 0	$ 0
Direct Costs	**$39,840**	**$49,840**	**$59,840**
Gross Profit (per 100 cows)	**$22,285**	**$27,410**	**$32,535**

Least farmer control: To keep the arithmetic simple, vary only the factor least in your control.

Calculating bull cost per 100 cows per year

Purchase @ $1,500 Sell @ $1,000 Life 5 years	Depreciation $100
Interest 10% $\frac{\$1,500 + \$1.000}{2} \times 10\% = \$125^*$	$125
Veterinarian and feed $50	$ 50
Miscellaneous	$ 25
	$300

1 bull to 25 cows $\frac{\$300}{25} = \$12/\text{cow}$

*If you buy a bull for $1,500 and sell it five years later for $1,000 then you take the first value and add the last value and divide by 2 to get the average between $1,500 and $1,000 (which is $1,250 on average tied up in that bull each year over five years) and 10% of that in interest is $125 per year.

Figure 1-4. Gross profit analysis prior to committing to an enterprise. This example looks at gross profit per 100 cows using poor, average, and good prices for sales and replacement costs.

could use the same method to find out instantly how changes in supplemental feed prices, labor policy, weaning percentages, average gain, and other factors would affect your gross profit.

Comparing the Use of Assets

Gross profit analysis allows you to estimate the effect of changing an asset from one use to another. This could be land, machinery, buildings, cash, or whatever. In table 1-1 we dissect the question of whether to use a piece of bot-

tomland for alfalfa or as pasture for stockers. The numbers are simply pulled out of a hat to show the technique and should not be viewed according to their plausibility. The analysis forces you to ask (and answer) many important questions:

- Which choice is most risky?
- Would your own machinery make hay cheaper?
- Could you raise your own stockers?
- Could you finance stockers yourself?

Table 1-1. Alfalfa or stockers?

Alfalfa—100 acres			Stockers—100 @ 1 steer/acre		
Income			**Income**		
3.5 ton/acre @ $125/ton	$43,750		750 lb steers @ $1/lb	$75,000	
Direct Expenses			**Direct Expenses**		
Custom cut, bale, and stack @ $75/ton	26,250		400 lb steers @$1.40/lb	56,000	
Seed and planting $6,000/5 years	1,200		Salt $ minerals, $1/head/month	750	
Fertilizer, herbicide	1,900		Shipping	2,000	
Application costs	1,050		Fence and miscellaneous	500	
			Interest	1,500	
Total direct expenses	$30,400		Total direct expenses	$60,750	
Gross profit	$13,350		Gross profit	$14,250	
Gross profit/acre	**$133.50**		**Gross profit/acre**	**$142.50**	

- Could you increase the stocking rate?
- Could you increase yield?
- Could you cut chemical costs?
- Which gives you the most flexibility in the future?

Gross Profit per Unit

If the mathematics of reducing gross profit to a unit of something tends to slip your mind, remember the following: $/head, $/acre, miles/hour, pounds/square inch, and so on are written like fractions, because they are. The top gets divided by the bottom.

If 100 cows produce a gross profit of $24,000, then

$$\text{Gross profit/cow} = \tfrac{\$24,000}{100} = \$240.$$

If the herd uses 800 acres, then

$$\text{Gross profit/acre} = \tfrac{\$24,000}{800} = \$30.$$

If the herd would cost $50,000 to replace, then

$$\text{Gross profit/\$ invested} = \tfrac{\$24,000}{\$50,000} = \$0.48.$$

That is commonly expressed as "forty-eight cents on the dollar" or 48 percent.

The Dangers of Gross Profit Analysis

Gross profit analysis tells you what each enterprise contributes toward fixed costs, or overhead, and possible profit. Only the final financial plan, however, really tells you if the enterprises you have chosen will together pay all the fixed costs and actually return a profit. In any case, you must develop a full financial plan to make sure that all your fixed costs are in fact paid. Like any other technical gimmick, gross profit analysis does not replace common sense.

Especially in crop farming, a particular crop may promise a high gross profit but fail all the other context checks because of its use, its nature, or its associated technology. The farmer whose hand calculator proves that cotton yields more dollars per acre than alfalfa on the strength of a gross profit analysis could easily destroy her land by monocropping cotton year in and year out. The family that made a killing in strawberries this year should not necessarily cut down an apple orchard to double the berry crop.

Diversity of enterprises is generally wise as a hedge against changing conditions in the marketplace. Biological diversity is vitally important to sustainable farming and ranching. For both these reasons, relying on a gross profit analysis alone can be dangerous. The use of the remaining six context checks avoids the trap of cash returns at the expense of true biological capital and, eventually, quality of life.

The simple form of the gross profit analysis described here has one other obvious limitation as well. It does not necessarily account for time and the availability of some resources. You cannot, for instance, schedule crops or other operations that demand more labor or equipment than you have on a given day, and this could become a limiting factor.

As you acquire more skill in performing the gross profit analysis, you might find some benefit in analyzing various combinations of crop enterprises as a single enterprise. Likewise, you might analyze different phases of, say, cattle production on the ranch—breeding to calving, or growing calves to eighteen-month-old steers, versus not breeding but bringing in young stock, and so on.

Management Tips and Good Ideas

Over the years, Holistic Management practitioners have developed a number of good ideas that have proven helpful in creating and implementing their plans. Here are four that have been put to use in a variety of situations and countries.

Simple Appreciation: How to Handle Emergencies

Even with perfect planning, if it were possible, things will go wrong and emergencies will arise—in your personal life or in any aspect of your operation. Because one cannot account for these situations in routine decision making and planning, we needed a way to address them that takes into account the fact that people are stressed in emergencies, and this clouds their judgment. Allan Savory found a practical way to overcome this dilemma by referring to a profession that has dealt with it for centuries.

Soldiers have long been taught to substitute routines for panic, so when bullets fly they duck and take aim before fear makes them stand up and run. It seems they also learn a routine for thinking when no routine applies. The technique, referred to as a simple appreciation, is standard in British military training.

First, take a deep breath to ease your panic. Then, before you engage in endless and futile discussions, take

out a notepad and start writing, using the following four headings:

1. *Aim:* Define your objective in one sentence. What result do you want? If you need more than one sentence, you have not focused enough. Try again until you have your aim very clear.

2. *Factors:* List everything, good or bad, that has a bearing on the situation. Pour your concerns onto the paper without worrying about any particular order or neatness. Don't be judgmental, and do it quickly. You won't be looking at what you write again. This step is just intended to get what's in your head out in the open.

3. *Courses:* Outline any courses open to you that would achieve your aim. You will seldom have more than three.

4. *Plan:* After some thought, pick the course you feel is most likely to achieve your aim. Then plan whatever it is you need to do to bring about the desired outcome.

Forcing yourself to go through these steps will clear your head and focus your imagination. You will be amazed at the way this simple appreciation brings forth clarity, which days of discussion and worry seldom do.

Decision-Making Matrix: Tracking Your Decisions

Changing the way you make decisions does not come naturally or easily. At the first sign of stress, it's all too easy to revert to old habits. A number of people have found a way around that by using a decision matrix, similar to the one shown in figure 1-5, to help discipline themselves.

When you create your financial plan for the year, you are making decisions about a number of actions you will implement during the year. Having a record of how you made those decisions will come in handy later on. For instance, some of the decisions you made may have been based on assumptions that no longer apply months later when it's time to actually implement the decision. If you

Context Checking Matrix

Proposed action/decision: _____   Date: _____

CONTEXT CHECKS	REMARKS	✓	X
Cause and Effect Does this action address the root cause of the problem?			
Weak Link—Social Could this action, due to prevailing attitudes or beliefs, create a weak link between us and those whose support we need?			
Weak Link—Biological Does this action address the weakest point in the life cycle of this organism?			
Weak Link—Financial Does this action strengthen the weakest link in the chain of production? (Resource, product, money conversion)			
Marginal Reaction *(comparing two or more actions that have passed the other checks)* Which action provides the greatest return toward the goal for each additional unit of the time and money spent?			
Gross Profit Analysis *(comparing two or more enterprises)* Which enterprises contribute the most to covering the overheads of the business?			
Energy/Money—Source Is the energy or money to be used in this action derived from the most appropriate source in terms our holistic context?			
Energy/Money—Use Will the way in which the energy or money is used be in line with our holistic context?			
Sustainability If we take this action will it lead toward or away from the future resource base described in our holistic context?			
Gut Feel How do we feel about this action now? Will it lead to the quality of life we desire? Will it adversely affect the lives of others?			
Monitoring Criteria:			

Figure 1-5. A useful tool for disciplining yourself to check your decisions as you create your Holistic Financial Plan. It provides a record you can refer to when memories grow fuzzy on the details. Indicate pass, fail, don't know, or not applicable in the column for each check. Use the "Remarks" column to note any information you feel may be needed when you later review the decision made.

have recorded your decision and the results of the checking, a quick recheck may warn you that a change is needed prior to implementation.

A good decision-making matrix also includes space to record the criteria you will monitor to ensure that the decision was the right one. It reminds you to make a conscious effort to monitor your progress, and it helps minimize the chance of argument when management team members have different recollections about what was said.

Brainstorming and Brainwriting: Generating New Ideas

Brainstorming helps open the minds of planners to new ideas. "Serious" people sometimes shy away from it, but research has shown that the most original and fruitful thinking occurs during moments of humor and playful competition. Since the very phrase "financial planning" tends to produce the opposite mood, an exercise such as brainstorming can help. But to be most effective, a brainstorming session also needs to include some "brainwriting," where participants reflect individually and record their ideas on paper following the stimulation of the group interaction. Some of the best ideas often emerge at this point.

The rules are simple:

1. Gather everyone who might have an interest in the task at hand—such as possible new enterprises, or ways to cut expenses—and some people who don't.

2. Divide the people into groups of five or more (eight to ten maximum), and let each group appoint a recorder.

3. Announce a timed competition (about three minutes) for the longest list of ways to solve a lighthearted problem. Any idea will do, no matter how crazy. No judgment will be made. Only the number counts. Pour out ideas. Do not stop to talk or discuss.

   Here are some typical challenges for the opening competitions:

   * What uses could you find for . . . (some simple object)?

   * How would you get an interview with . . . ?

   * How might you deliver a proposal of marriage to . . . ?

4. Start the timer and at the end of the time read the lists and give a humorous prize to the team that came up with the most ideas.

5. Now put the serious problem on the table and let the groups compete in the same way (for ten minutes), with members listing any idea or solution that pops into their head, no matter how fantastic.

6. Appoint a facilitator who doesn't participate but monitors each group to ensure members observe the following rules:

   * *Do not allow anyone to judge any idea.*

   * *Do not allow anyone to dominate the conversation.*

   * *If a silence develops don't allow anyone to talk.* As soon as someone talks to break the awkward silence, creativity drops. Some of the best ideas suddenly flow after a period of silence.

   * *Allow cheating.* If someone at one table happens to hear an idea from another table, it can start a fresh run of ideas at the cheating table.

7. Next, have members work on their own for another ten minutes—*no talking allowed*—to come up with additional ideas or to refine those already presented, and record them on a separate piece of paper.

8. From these final lists (both group and individual) of ideas pick out the ones that have potential and develop them.

To emphasize the point that open speculation often produces more than reworking old ground, Allan Savory tells the following story:

While I was touring a ranch in Africa, my host took me to a particularly rough and inaccessible section that he said had no use whatever. It was steep, waterless, full of predators, and a thorough headache to a stockman, but it was beautiful. "Why not put some cabins down there where couples from the city could hide out for a little romance?" I asked.

The rancher, a man of righteous constitution, took great exception to the hint of impropriety, but the notion that natural beauty and tranquility had cash value in themselves stuck. He built the cabins, and within a few years they produced more revenue than the rest of his operation combined. The moral is that pounds of beef per acre or hectare may be only one of a million ways to assess the potential of land and the value of its production.

In a similar vein it might be remembered that when Ebenezer Bryce first looked down on Bryce Canyon in Utah, he could only remark: "That's a hell of a place to lose a sheep." Yet how many millions of Americans have driven how many millions of miles for the same view?

Net Managerial Income: Rewarding Managerial Effectiveness

Paying employees a bonus based on production, income, or profit is an old idea. Unfortunately, its potential for inspiring people to make their work more productive usually remains unfulfilled—the big bottom-line figure is so remote from daily decision making that most workers never really know how they fit in. All too often, if managers push production or income, they also push up costs; or if profit is the criterion for a bonus, it can raise issues of mistrust as to how profit is determined. And there is ample research to show that while bonuses produce "movement," they often do not provide motivation.

If you are intent on providing a bonus, one technique you might consider is a bonus based on a percentage of what we call *net managerial income* (NMI). This is a true measure of managerial effectiveness, and it places pressure on managers to both increase income and decrease costs. It is determined only from those aspects of income production and cost control that are entirely in the hands of managers and not influenced by owner whims or perks. Additionally, it is true "open-book management," in that all parties have access to the figures.

When managers participate in financial results, NMI provides a fair measure on which to base their compensation. NMI is the difference between gross income and gross expenses, taking into account only income and expenses under the direct control of a management sector. This could apply to a foreman who has total responsibility for a band of sheep. It could be the managers of farming and livestock divisions in a large operation.

NMI accounting might also recognize the frequently made distinction between "owner control" and "management control." In this case, for example, the owner oversees such matters as mineral leases, real estate transactions, and futures trading, while a separate management team runs agricultural production.

NMI is the simplest index of managerial effectiveness. Setting up columns on your planning chart (described further on) to show it graphically month by month can provide a major staff incentive—especially if NMI is connected to financial rewards and promotions.

NMI has obvious advantages because it creates incentives for managers to cut costs as well as raise income, and it keeps them aware of where the costs are. Unfortunately, though, there is no universal formula for relating NMI to actual staff payments because not all submanagers control equal shares of income and cost amounts. For the manager of a cattle enterprise, for instance, the main cost may be fixed land payments over which she has no control. If you take land payments out of the cattle account, however, you will show an unrealistic net income. In contrast, the manager of a chicken facility may have great control over the big expense of that enterprise—the feed.

You can compensate for such inequities, but often the exercise will cause you to redefine jobs. If, for instance, the person who manages field crops has no responsibility for machinery costs and maintenance, you might want to make that part of the NMI account and grant the authority necessary for controlling those costs. Such reorganization always runs the risk of upsetting entrenched interests. Be sensitive, proceed openly, and make every effort to build trust in advance.

NMI may show what appears to be a very high "profit" because costs not directly under control of the managers are ignored to arrive at NMI. However, those costs are still real and have to be paid before there is any real profit to share. Thus, if you are giving a bonus derived from NMI, it becomes essential to use some formula to account for those "owner costs" or fixed costs not under managerial control before a percentage of what remains is distributed. You could, for instance, determine the bonuses by first multiplying total wages and salaries by two or three and subtracting the total from the "profit" figure. Then you would decide what percentage to distribute. If something like this is not done, the percentage of NMI that could be shared would need to be very small, or it would break the owner. It is better to make an allowance for those ignored costs and then provide for a larger percentage of remaining NMI.

Planning Forms

Before combining all your ideas into a single grand plan, you must work out the details of each aspect in terms of income, labor, quantities, expense, and, most of all, time. The Savory Institute has developed forms to help you do this. Time figures critically in all decisions. Workloads that coincide may require more labor. Expenses that come before your products are marketed may require credit. Construction not completed on time can cost you an entire crop.

You will need four different forms to both develop and monitor your plan. A sample of each, reduced in size, is included in appendix 2, as follows:

- **Worksheets:** These are the standard worksheets used to record the detail of every income and expense item (see figs. 1-6, 1-7, and 1-8).

- **Annual Income and Expense Plan:** This multicolumned spreadsheet summarizes the figures contained on all the worksheets, resulting in a master plan that relates time, income, expenses, cash, bulk purchase consumable items, and credit, and includes a planned profit. Printed versions are useful for assembling data, but computer spreadsheets can easily be set up to handle the endless modifications and trial plans called for in the planning process. This same form is used as you monitor the plan's implementation.

- **Livestock Production Worksheet:** This is really a glorified worksheet for detailing livestock breeding operations.

- **Control Sheets:** These forms are for reporting items that run counter to plan and recording decisions on corrective actions and who is responsible for implementing them.

Any other forms or scraps of paper should be avoided. They add to confusion and stress, especially in the most difficult part of planning—control. All of these forms are available from the Savory Institute website (http://savory.global).

In the following sections, we elaborate on the uses of the standard worksheet and how to use the Livestock Production Worksheet. The other two forms are described in more detail in the next section, "Creating Your Plan."

Standard Worksheet

The standard worksheet has columns for the months of the year (you can start with any month to encompass the natural cycle of a task) and rows for different categories of expense, work, income, decisions, or whatever. The possible uses are far too varied to illustrate here. Your whole operation is likely to require dozens of worksheets and many drafts. You can use final drafts not only for assembling the master plan but also as general work schedules, as inventory and budget guides, and as the basis for all ongoing monitoring and controlling. The examples in figures 1-6, 1-7, and 1-8 show something of the standard worksheet's versatility.

Figure 1-6, for example, shows the biological year for a commercial cattle herd, which is useful in the following ways:

- It helps in designing a culling policy or when changing herd size.
- It shows animals in each class for figuring supplement quantities, forage needs, and the like.
- It helps you to plan sales policy, holding late calves, breeding back open heifers, and so forth.

Livestock Production Worksheet

When completed, the Livestock Production Worksheet tells you when your expenses and income will occur and predicts the effect of breeding rates, culling policies, and the like over a long period. For example, you can tell at

Biological Year of a Cow Herd

Date __2018__   **WORKSHEET**   Planning Sheet
Column Reference _____   Worksheet No. __3__

	Jan	Feb	March	April	May	June	July	August	Sept	Oct	Nov	Dec	
Seasonal Year	←—Calving—→			←Bulling—→					Weaning				
Mature Cows	←3+ years of age→			←Bulling→									
Heifers 2	←36 mo Calving→			←2nd Bulling→ 27 mo		30 mo	31 mo	32 mo	33 mo	34 mo	35 mo		
Heifers 1	←24 mo Calving→			←1st Bulling→ 15 mo		18 mo	19 mo	20 mo	21 mo	22 mo	23 mo		
Heifers	12 mo	13 mo	14 mo							9 mo	10 mo	11 mo	
Steers	12 mo	13 mo and sold somewhere from here to 18 mo depending on market								9 mo	10 mo	11 mo	
Culls										For sale anywhere from Oct on			
Bulls				—— HAVE BULLS ALL YEAR LONG ——									
Months of age Transfers													
Total													

Figure 1-6. The top line (seasonal year) shows major events in the biological year—calving, weaning, and such. The arrows show when animals progress from one class to another or go to sale.

Alfalfa Production and Sales Plan

Date __2018__   **WORKSHEET**   Planning Sheet
Column Reference _____   Worksheet No. __5__

	Jan	Feb	March	April	May	June	July	August	Sept	Oct	Nov	Dec	
160 Acres Alfalfa											960 tons kept use on the ranch		
Harvest av. 3 ton/acre						480 tons	480 tons	480 tons				1,440 tons	
Sales		sell 120 tons				sell 120 tons	sell 120 tons					sell 120 tons	
Estimate Price		$100/ton				$80/ton	$80/ton					$90/ton	
Cash Income		$12,000				$9,600	$9,600					$10,800	$42,000
Total													

Figure 1-7. This alfalfa production and sales schedule uses a strategy for spreading income throughout the year. Later in the planning this could be modified as the cash-flow situation becomes evident. A total of 1,440 tons of alfalfa is raised on 160 acres of land. The alfalfa is baled in June, July, and August. Two-thirds, or 960 tons, are kept to be fed on the ranch. One-third, or 480 tons, is sold—120 tons in February and another 120 tons in June, July, and December, at varying prices.

Fuel Use and Purchase Projections

Date __2018__   **WORKSHEET**   Planning Sheet Column Reference __Fuel__   Worksheet No. __14__

Seasonal Year	Jan	Feb	March	April	May	June	July	August	Sept	Oct	Nov	Dec	Item Totals
Seasonal Year	← Feeding and Calving →			← Bulling →					Weaning	Sales	← Feeding →		**Item Totals**
Manager's Suburban	200m	250m	250m	250m	250m	250m	250m	250m	250m	400m	100m	100m	2,800 miles
Suburban Off Ranch			4,000m					2,000m					6,000 miles
Jim Jones Pickup	200m	200m	150m	50m	50m	50m	50m	50m	50m	50m	200m	200m	1,300 miles
Jones Motorbike			100m	100m	300m	300m	300m	300m	300m	300m			2,000 miles
Jake Moncho Pickup	350m	350m	250m	250m	60m	60m	60m	60m	60m	60m	350m	350m	2,260 miles
Moncho Motorbike			150m	150m	400m	400m	400m	400m	400m	400m			2,700 miles
John Smith Mechanic P.U.	100m	100m	100m	300m	300m	300m	300m	500m	500m	500m	100m	100m	3,200 miles
Pickups miles/ gals @ 12 mpg	850m/ 71g	900m/ 75g	750m/ 63g	850m/ 71g	660m/ 55g	660m/ 55g	660m/ 55g	860m/ 72g	860m/ 72g	1010m/ 84g	750m/ 63g	750m/ 63g	9,560m/799g
Purchases off ranch	4,000 miles estimated		$866			2,000 miles estimated		$434					$1,300
Motorbikes miles/ gals @ 90 mpg			250m/ 3g	250m/ 3g	700m/ 8g	700m/ 8g	700m/ 8g	700m/ 8g	700m/ 8g	700m/ 8g			4,700m/52 g
Monthly gas consumption	71g	75g	66g	74g	63g	63g	63g	80g	80g	92g	63g	63g	To plan sheet by month
500 gals bulk purchase gas		$1,200						$1,200 approximately 147 gallons on hand at year's end					$2,400
Monthly Totals	$1,200	0	$866	0	0	0	$1,200	$434	0	0	0	0	$3,700

Figure 1-8. This worksheet shows fuel purchases for ranch vehicles. It projects consumption by various employees, inventories of fuel bought in bulk, cost and timing of bulk purchases, and monthly off-ranch purchases. This illustrates the best way of arriving realistically at both consumption figures and bulk-buying cash requirements. It will require two separate columns on the final spreadsheet—one for consumption and one for expense figures.

once what effect a five percent drop in conception rate this year will have on your herd size in three years.

Each Livestock Production Worksheet supports two separate columns on the master plan spreadsheet: an income column (to track livestock sales) and an expense column (to track livestock purchases). All other livestock expenses such as feed, medication, and so on are planned separately on standard worksheets.

To develop production plans for each of your livestock enterprises, do the following:

1. Use a separate Livestock Production Worksheet for each enterprise (commercial cattle, registered cattle, llamas, sheep, etc.). Are you sure, from a gross profit analysis, that the production policy for each enterprise is sound?

2. For each Livestock Production Worksheet draw up a standard worksheet with the biological year (calving, weaning, culling, breeding, etc.) clearly laid out, similar to the work sheet in figure 1-6. This helps ensure a clear picture so that you have no confusion on when animals change age, class, and so on. (Note: It's easiest if you start with the youngest animal class and follow it through its life.)

3. Record in column A on the Livestock Production Worksheet (see figure 1-9) the different classes of livestock that will be on the ranch or farm at any time during the year—for example, bulls (old and young), cows (mature), heifers (bred), heifers (open), male

Livestock Production Worksheet, 1-Year Plan

Savory

RANCH __River Bend__

ENTERPRISE __Cattle__

REMARKS __One year analysis__

LIVESTOCK PRODU

January 2019   YEAR OR MONTHS __April/May__

A CLASS OF STOCK (Cows, Heifers, Calves, Bulls, etc.)	B BIRTH % EST.	C OPEN NO.	D AGE	E BIRTHS	F MONTH	G BUY	H MONTH	CLASS TRANSFERS IN	OUT	I DEATH	J %	K SALE	L MONTH	M CLOSE AND OPEN NO.	N AGE	E BIRTHS	F MONTH	G BUY
1. Bulls		20				5	2		1	0				25				
2. Cows	90%	500						200	2					700				
3. Heifers – 2	70%	200						225	3 200	0				225				
4. Heifers – 1	90%	225						240	4 225					240				
5. Heifers		240							5 240	0				0				
6. F. Calves		0		396	2				6	0				396	4			
7. M. Calves		0		396	2				7	0				396	4			
8.									8									
9.									9									
10.									10									
11.									11									
12.									12									
13. TOTAL HEAD		1185		792										1982				

NOTE: RECORD YOUR ESTIMATES OF PERCENTAGE ACTUAL BIRTHS FOR THE VARIOUS AGE CLASSES OF BRED FEMALES IN THEIR ROWS IN CO

	Bulls		Cows		Heif-2		Heif-1		Heifers		M. Calves		Bulls	
14. CLASS OF STOCK	Bulls		Cows		Heif-2		Heif-1		Heifers		M. Calves		Bulls	
15. NUMBER SOLD/MONTH OF SALE	5	10	138	11	66	11	24	11	38	11	380	11		
16.														
17. AVERAGE LIVE WEIGHT	1500		950		900		800		450		550			
18. MEAT PRICE PER LB.	8.0		0.75		1.20		1.25		1.40		1.50			
19. INCOME PER ANIMAL	1200		713		1080		1000		630		825			
20. WOOL/HAIR WEIGHT/MONTH OF SALE														
21. WOOL/HAIR PRICE/MONTH OF SALE														
22. WOOL/HAIR INCOME														
23. PLANNED GROSS INCOME	6000	10	98394	11	71280	11	24000	11	23940	11	313500	11		
24. NUMBER PLANNED TO BUY													5	
25. ESTIMATED PRICE/ANIMAL													2500	
26. TOTAL COST AND MONTH													12,500	12

Figure 1-9. This worksheet tracks a herd for one year divided into three four-month periods according to the biological year described in the worksheet in figure 1-6. It reflects a policy of culling twenty percent of mature cows, ten percent of female calves, and all unproductive heifers (ten percent of Heifers-1 and thirty percent of Heifers-2). Study the numbers with a calculator in hand until you get the hang of how the numbers develop. Note that rows 14–23 are income and rows 24–26 are expenses.

and female calves, and steers. Keep the categories in the same order as they appear on your biological year worksheet to reduce the chance of errors.

4. Decide whether you are using the worksheet for one year or for more than one year. The form has three similar sections across it. One of these can be used for one year and

DUCTION WORKSHEET

DATE OF PLAN ___June 2018___

YEAR OR MONTHS Aug/Sept									YEAR OR MONTHS December										
G BUY	H MONTH	CLASS TRANSFERS IN	OUT	I DEATH	J %	K SALE	L MONTH	M CLOSE AND OPEN NO.	N AGE	E BIRTHS	F M	G BUY	H MONTH	CLASS TRANSFERS IN	OUT	I DEATH	J %	K SALE	L MONTH	M CLOSE NO.	N AGE
		1		0				25						1				5	10	20	
		2		14	2			686						2				138	11	548	
		3		5	2			220						3				66	11	154	
		4		5	2			235						4				24	11	211	
		5		0				0						380 5				38	11	342	
		6		16	4			380	8					6	380						0	
		7		16	4			380	8					7				380	11	0	
		8												8							
		9												9							
		10												10							
		11												11							
		12												12							
				56				1926										651		1275	

IN COLUMN (B) ABOVE.

ANALYSIS OF PLANNED SALES AND PURCHASES

the other two left blank. Three sections, or one whole worksheet, can be used for one year divided into three four-month periods, as shown in figure 1-9. Alternatively, when you are estimating the rate of growth of a herd over a few years, one worksheet can be used to reflect three years, as shown in figure 1-10.

5. Record the number of animals in each class at the start of the financial year in column C. Look at the biological year worksheet for clarity. Record the age in months for each class in column D. If the animals are mature, you can record 36+. For a group of young animals, you can use the average age.

Savory

Livestock Production Worksheet, 3-Year Plan

RANCH __River Bend__

ENTERPRISE __Cattle__

REMARKS __3 year plot to see rate of herd growth__

LIVESTOCK PRODUC

YEAR OR MONTHS __2019__

A CLASS OF STOCK (Cows, Heifers, Calves, Bulls, etc.)	B BIRTH % EST.	C OPEN NO.	D AGE	E BIRTHS	F MONTH	G BUY	H MONTH	CLASS TRANSFERS IN		CLASS TRANSFERS OUT	I DEATH	J %	K SALE	L MONTH	M CLOSE AND OPEN NO.	N AGE	E BIRTHS	F MONTH	G BUY
1. Bulls		20				5	2		1		0		5	10	20				7
2. Cows	90%	500						200	2		14	2	138	11	548				
3. Heifers – 2	70%	200						225	3	200	5	2	66	11	154	36			
4. Heifers – 1	90%	225						240	4	225	5	2	24	11	211	24			
5. Heifers		240						380	5	240	0		38	11	342	12			
6. F. Calves		0		396	2				6	380	16	4			0		395	2	
7. M. Calves		0		396	2				7		16	4	380	11	0		395	2	
8.									8										
9.									9										
10.									10										
11.									11										
12.									12										
13. TOTAL HEAD		1185		792		5					56		651		1275		790		7

NOTE: RECORD YOUR ESTIMATES OF PERCENTAGE ACTUAL BIRTHS FOR THE VARIOUS AGE CLASSES OF BRED FEMALES IN THEIR ROWS IN CO

14. CLASS OF STOCK							
15. NUMBER SOLD/MONTH OF SALE							
16.							
17. AVERAGE LIVE WEIGHT							
18. MEAT PRICE PER LB.							
19. INCOME PER ANIMAL							
20. WOOL/HAIR WEIGHT/MONTH OF SALE							
21. WOOL/HAIR PRICE/MONTH OF SALE							
22. WOOL/HAIR INCOME							
23. PLANNED GROSS INCOME							
24. NUMBER PLANNED TO BUY							
25. ESTIMATED PRICE/ANIMAL							
26. TOTAL COST AND MONTH							

Figure 1-10. This worksheet tracks a herd for three years, enabling you to work out how fast your herd will grow without buying in cows.

6. Decide on the actual birth percentages you anticipate for all bred females of different age groups. Record these percentages in column B opposite the appropriate females in column A.

7. Estimate the mortality percentage you anticipate with each class of stock and record it in column J. If you are using the worksheet for three years, do this for each year in column J. If you are using the worksheet for one year only, use column J in the middle of the year for simplicity and record mortalities as though they all occurred midyear.

8. Next, using the percentage birth estimates (column B), calculate the births expected in the financial year from each group of females

ꓒUCTION WORKSHEET

DATE OF PLAN __June 2018__

YEAR OR MONTHS __2020__   YEAR OR MONTHS __2021__

G BUY	H MONTH	CLASS TRANSFERS IN		CLASS TRANSFERS OUT	I DEATH	J %	K SALE	L MONTH	M CLOSE AND OPEN NO.	N AGE	E BIRTHS	F M	G BUY	H MONTH	CLASS TRANSFERS IN		CLASS TRANSFERS OUT	I DEATH	J %	K SALE	L MONTH	M CLOSE NO.	N AGE
7	2		1		0		5	10	22				7	2		1		0		5	10	24	
		152	2		14	2	109	11	577						142	2		15	2	114	11	590	
		207	3	152	5	2	62	11	142	36					295	3	142	6	2	88	11	201	
		335	4	207	7	2	34	11	298	24					334	4	295	7	2	33	11	297	
		379	5	335	8	2	38	11	340	12					429	5	334	9	2	43	11	383	
			6	379	16	4			0		447	2				6	429	18	4			0	
			7		16	4	379	11	0		447	2				7		18	4	429	11	0	
			8													8							
			9													9							
			10													10							
			11													11							
			12													12							
7					66		627		1379		894		7					73		712		1495	

ꓲN COLUMN (B) ABOVE.

ANALYSIS OF PLANNED SALES AND PURCHASES

bred. Check the biological year worksheet for each group. They may be bred this year but produce no offspring until the next financial year. They may not have been bred last year. When you estimate births from each group, you can generally, for simplicity, group animals born and divide by two to provide an estimate of male and female calves, lambs, and so forth. Record estimated births in col-umn E opposite the rows for male and female calves, lambs, and so on. Record the average month of birth in column F.

9. Plan any livestock purchases that will be necessary during the year and record them in column G. Record the planned month of purchase in column H. Do this in the appropriate row for any class of livestock to be purchased.

10. Next, take each class of livestock again. It is usually easiest to start with the youngest and follow them through the year (refer to your biological worksheet).

    • Note their age at the beginning, add any purchases, and calculate any deaths (from the mortality estimates in column J). Record the estimated number of deaths in column I.

    • Decide if any of that class of animal will be sold, and if so record the number in column K. Record the planned month of sale in column L.

    • Check your biological year worksheet to see if the group will change in age class during the year; if so, record this change in the shaded class transfer columns. Record the number passing OUT of that class in the appropriate row. Record the same number IN, in the row for the class of stock that they will become. The OUT number must always equal the IN number. Remember, if you have already taken off deaths for those animals, not to do so again in the class you have transferred them to.

    • After following each group through the year, record the number you will close with in column M. Record their average age in months in column N. This closing column also becomes the opening column for the next year planned. If you are using one worksheet for a year in three four-month periods, as shown in figure 1-9, close each four-month period in this manner.

11. Check your plans by adding all the figures for livestock you started the year with. Record the total in row 13 ("Total Head") in column C. Add and record totals for columns E, G, I, K, and M in row 13.

    Take the opening total (column C) and add to it the totals of columns E and G. Sub-

tract from this the totals of columns I and K, and the answer should equal the total under column M. If not, you have an error to find! In the case of the shaded transfer columns, merely check that IN and OUT totals are equal. Otherwise, leave them out of the totals in row 13.

12. Next, analyze the sales. In row 14 ("Class of Stock"), record in one of the columns available each type of animal that will be sold. Next, look in column K and transfer the number for sale in each class to row 15, together with the month of sale (from column L). Note that you have a spare row (16) in case more than one group of a particular class is sold. Do the following for each class of animal to be sold:

    • Estimate the weights and note them in row 17.

    • Estimate the price per pound and note it in row 18.

    • Calculate the income per animal and note it in row 19.

    • Add wool, hair weights, and prices, if applicable, and note them in rows 20–22.

    • Calculate the planned gross income for each class of livestock and record it in row 23. These are the final planned INCOME figures that should be transferred to the main spreadsheet.

13. Analyze the planned purchases. First, record in row 24 the classes of livestock to be bought, using as many of the available columns as necessary. Next, estimate the price per animal and record it in row 25. Calculate the total cost and record it in row 26. To the right of that, record the month of planned purchase taken from column H above. These are the final planned livestock purchase (EXPENSE) figures, which should be transferred to the main spreadsheet columns dealing with the particular enterprise.

Creating Your Plan

WHAT FOLLOWS IS A PROCEDURE that originated in a series of steps Allan Savory worked out for creating and using a total farm or ranch plan that accounted for both money and biological capital. It differs fundamentally from the conventional cash flow budgeting that agriculture adopted from other industries, which tends to determine priorities in terms of cash and cash return only.

Planning the Planning: Laying the Foundation

Most cost accounting procedures treat all capital items as "wealth generating" so long as they can be shown to mathematically return enough revenue to pay for themselves and provide an acceptable margin. This approach all too often leads to failure though, because wealth-generating potential also includes such noncash items as soil productivity, water quality, and training.

The common practice also fails to rank investments in relationship to larger goals, or what today is referred to as the triple bottom line (social, environmental, and economic accountability). Again, anything goes as long as it returns enough dollars. Here, we try to discover what will return enough in terms of goals that may not be quantifiable, say, maintaining a diversity of species, and to also align planned actions with a holistic context that reflects our deepest values.

This makes a particular difference in funding capital investments. Instead of putting your money into many projects simultaneously, you will probably find yourself tackling one at a time but more intensely, as not many are likely to pass the context checks in a given year.

The planning process outlined here represents many years of evolution and the experience of practical farmers and ranchers who have little tolerance for unproductive paperwork. If you find the description daunting, rest assured that it is more straightforward to *do* than to read. It will work as well for small operations as large ones and brings the greatest return in times of crisis.

Mental attitude is far more important than financial resources. You have to carry through, be creative but tough minded, and not fudge to make things look rosier than you know they are. For starters, however, don't think of the task as a drudgery of numbers and balances. Think of yourself as a conductor about to direct a symphony. You are about to exercise that kind of mastery over your own life and a stretch of land.

If you can acquire computer software to handle the main spreadsheet, do it. Any sophisticated spreadsheet program, such as MS Excel, will work. Using a computer eliminates the toil of balancing columns and erasing yards of numbers every time you change a detail of the plan. Tedious hours of calculating, locating errors, and redoing work by pencil and paper can be discouraging and will drain your creativity. But consider using the paper worksheets and spreadsheets the first time through to make sure you fully understand the process before modifying it to suit your software.

Creating Ownership: Who Should Be Involved?

Don't try to plan alone. Select a team that is as inclusive as possible without being unwieldy. There are several reasons for this strategy:

- People who become involved with projects at the beginning will care about them and see them through to the end.

- If people are going to be held responsible for generating income or holding down expenses, they should be allowed to come up with the figures that are directly under their control.

- When hard choices must be made, morale will survive much better if everybody understands why and has a chance to work on solutions.

- It takes a lot of work to prepare for good planning. Information gathering, inventories, gross profit analyses, time sequences—everything that requires a worksheet—might keep one person tied up year-round. A team, especially if it includes people intimately acquainted with field realities, gets through that work faster and better than any individual.

Overview and General Scheduling

If you haven't ever done it before, get organized to start planning as soon as possible. After that, you routinely start the process two to three months before the financial year begins. Altogether, the planning is not likely to take more than a week of your time, and most of that will be in hours snatched here and there for tracking down figures, thinking through the actions you might take to address a weak link, and so on. The bulk of your time will probably be spent on the first of the three phases that make up Holistic Financial Planning:

1. **Preliminary planning:** This phase is devoted to reviewing the current year's plan, gathering information and figures for the new plan, and running decisions through the context checks. It will require at least one lengthy team meeting to finalize the steps involved, even if only two of you are planning.

2. **Putting the plan on paper:** All the thinking you did in the preliminary phase, and all the decision checking and sorting of information you compiled, now comes into play as you put your plan on a spreadsheet. If you've done your preliminary planning well and researched the costs involved in the ideas you want to implement, you probably won't need more than a day or two, using a computer, to complete this phase.

3. **Implementing and monitoring the plan:** This phase is continuous and generally requires a few hours once a month throughout the year. All the figures you've planned must be monitored monthly to ensure that they remain on track, since events rarely turn out exactly as planned. If actual figures deviate from planned figures, take action immediately to get back on track. Good monitoring and controlling of your plan help avoid the need for replanning.

Schedule your planning sessions in advance so commitments and deadlines mean something. Take care not to allow too much time between sessions, or your momentum could suffer.

Your planning schedule should include the following:

- If this is your first time to plan, a social event, potluck, or barbecue where you will brainstorm new sources of income and/or cost-cutting strategies (see box 1-3). Keep the atmosphere light. Include families and children. Don't condemn any of the ideas that arise. Do this well before you start planning if you are brainstorming new sources of income—at least two or three months prior.

- Time to gather information. Alert team members to research or think through everything potentially relevant—gross profits, sales trends, equipment costs, inventories, and so on. It can be extremely frustrating to gather for a planning session on an isolated ranch or farm, only to discover that no one has a clue how much a mile of fence or a new tractor will cost.

Box 1-3. Optional Step: Brainstorming New Sources of Income

Most people tend to put themselves and their operations into a well-defined box ("I'm a cow-calf operator," "I'm a wheat farmer") and thus limit their possibilities for profit enhancement. Periodically, it pays to challenge your existing enterprises and the thinking behind them, because there may be other, far more profitable things you could be doing using the same resource base.

But don't overdo it. Few operations survive by chopping and changing enterprises every year. Any new enterprise involves a learning curve and will take a year or more to prove out, and in most cases, it will require a business plan that looks several years ahead. Brainstorming sessions held every three to five years are usually sufficient. Hold them as early in the year as you can—six to nine months prior to your fiscal year end isn't too early. You'll need the time to develop a business plan, if needed, and to track down the figures required for planning.

Follow the rules for brainstorming (*and brainwriting*) provided in the last section ("Mastering the Basics"), and you'll generate more ideas than you can ever use. Use the process described below to help you narrow down the list.

Narrowing Down Your Brainstorming List
Here's how to weed out inappropriate enterprises and determine which ones have the most potential:

First Cut: Drop any ideas that conflict with your values (as expressed in your quality of life statement). There are likely to be many if the brainstorming was free-flowing, fun, and irreverent.

Second Cut: Drop the ideas that are patently ridiculous, such as raising purple mice or engaging in the flea circus business. But be careful in labeling an idea ridiculous. Some of the best ideas are often initially ridiculed by people who are the most knowledgeable about the present situation.

Third Cut: Do a very rough gross profit analysis, as you will not have actual figures to use. Look at the remaining possibilities and, thinking broadly, ask what each might bring in as income over and above what it might cost to engage in that enterprise. You want to keep those enterprises that bring in the most income for the least cost because those will result in the most gross profit, which, after covering all costs, will finally lead to a net profit. At this point you're looking for major differences, not minor ones.

Fourth Cut: Use the rest of the context checks to help you assess whether the likely actions and inputs involved in each possible enterprise will move you closer to your goals while remaining aligned with your holistic context. Would the enterprise use significant energy? Where from and what type? Where would the money come from? If any of the tools or actions used in the enterprise would obviously lead you away from your goals or be misaligned with your holistic context, eliminate the enterprise for now.

Fifth Cut: You are likely to have only three to four of your original hundred or so ideas left by this cut. Now do a detailed gross profit analysis with well-researched figures, comparing each enterprise to the others and to your current enterprises, to see which ones really do contribute the most to covering your fixed costs and thus provide eventual profit.

- A half day or so, or several briefer sessions, to discuss and finalize the decisions arising from your preliminary planning.

- A day or two of uninterrupted time to create your plan on the spreadsheet. If you are not using a computer, you may need an additional day. Make this a planning retreat and get away from the farm or ranch if that is what it takes. Aim for reaching this point one month prior to your fiscal year end.

- A final half-day team meeting to make the adjustments needed to make the plan work. Try to work on a computer so you can instantaneously reckon the implications of change.

- Time to organize monitoring and control routines.
- Monthly team monitoring sessions, beginning in the new fiscal year.

Preliminary Planning: Taking a Hard Look

In this initial planning phase, you take a hard look at where you are now and where you need to go in the coming year relative to your goals and your holistic context. You will make a number of decisions that have a direct bearing on the income and expense figures you will enter onto worksheets. This preparation is essential. Forgoing it in favor of planning dollar figures directly on the spreadsheet leads to serious errors. You have four basic questions to answer:

1. Is there a logjam, and if so, how will you address it?

2. Are other factors adversely affecting the business as a whole?

3. Are current enterprises "profitable" and have you diversified your risk?

4. What is the weak link in each enterprise and how will you address it?

1. Is There a Logjam, and If So, How Will You Address It?

North American English acquired the term *logjam* from the practice of floating large quantities of logs down rivers to sawmills and ports. Logjams occurred whenever the flow hung up; however, the drivers discovered that these colossal piles of timber would generally move again if they could dislodge the one key log that had stopped the lot. In this case, what you want to ask yourself is whether anything might be blocking the business as a whole from making genuine progress toward its goals.

To get your answer, you need to step back from day-to-day operations and concerns to gain greater perspective, much as the logger does in attempting to clear a logjam. He climbs a hill or tree to see which of the thousands of logs floating down the river has caused the blockage. Then he removes the one responsible, and the rest flow. It is that one log you need to identify so you can progress once

more. And it can be difficult to identify and deal with as it commonly involves people in key positions and is related to their attitudes, mindset, trustworthiness, and so on.

Bear these points in mind:

- Sometimes it takes an outsider to see the logjam you cannot see because of your limited perspective. Finding a logjam and addressing it effectively is not always as easy, tidy, or obvious as it is in the logging analogy.

- In the beginning especially, the logjam may be tied to how the whole being managed was defined (e.g., a key decision maker may have been left out). Or it could be tied to how the holistic context was created (e.g., it was borrowed from someone else or created superficially and has little meaning for managers, and as a result it does not influence any actions).

Other common logjams bedeviling ranchers and farmers include the following:

- Government regulations that prevent ranchers from running enough animals or using adequate levels of animal impact

- Contractual agreements that lock a farmer into commodity production

- Lack of adequate water supplies, or the ability to deliver adequate amounts consistently to crops or livestock

- Lack of knowledge of viable alternatives to monocropping

- Lack of knowledge about or experience in mixing livestock and crops together

- Lack of knowledge about or experience in handling large herds, especially at breeding, calving, lambing, and kidding times

- Location of the farm or ranch (when producing specialty goods for urban markets)

- A brittle-environment ranch that is too small to be ecologically manageable (and will need to combine with others in order to run a large enough herd to keep the land healthy)

• Properties too large to be managed effectively by the number of people available or willing to manage (a common logjam in parts of Australia)

Many of these examples may in fact prove not to be logjams, as some progress, albeit slow, could still be made overall. In that case, they might appear in the "adverse factors" category (see point 2).

The actions you take to clear a logjam in your business receive the highest priority. And if they require money, it is allocated to them prior to any other expense.

Create a worksheet. Label it "Wealth-Generating Expenses" at the top. In the left-hand column, list any expenses you will incur to address a logjam (if found) and note "logjam" in parentheses beside them. If you have actual expense figures, include them (in pencil), as shown in figure 1-11. If you do not have actual figures, assign someone to research the costs. Pencil in that person's initials next to the item along with the date that information should be ready.

2. Are Other Factors Adversely Affecting the Business as a Whole?

In this case, you are mainly looking for things that reduce overall efficiency and productivity. They are less urgent than something that is blocking progress altogether, but still important. If not addressed, they could become a logjam at some point.

Savory

Wealth Generating Expenses

Date __November 2018__   **WORKSHEET**   Planning Sheet Column Reference _____ Worksheet No. __1__

	Jan	Feb	March	April	May	June	July	August	Sept	Oct	Nov	Dec	
WEALTH GENERATING													
Retreat/Coaching													
(Logjam)			1,500										
Office addition						10,000							
Computer		1,500											
Software		200											
Printer		500											
Total		2,200	1,500			10,000							

Figure 1-11. The logjam in this case was a lack of progress due to a hastily created holistic context in which no one felt ownership and consequently decision making had not changed. Thus $1,500 has been planned for a retreat away from the ranch where all members could participate with the help of a facilitator to coach them through the process. The remaining expenses are considered wealth generating, in this example, because the planning team determined that the lack of these items was leading to such inefficiency that income was being lost.

The following are common factors adversely affecting agricultural businesses:

- No time or funds allocated for vacations or to visit family
- Lack of capital to launch a promising new enterprise
- Lack of computer skills
- Lack of communication and cohesiveness as a family
- Lack of communications infrastructure to permit Internet access, fax transmissions, or email
- Working environment too toxic (with chemicals) or dangerous (with heavy machinery) to involve the children
- Lack of expertise and "connections" for marketing value-added products.

If any other factors are adversely affecting the business as a whole, now is a good time to recognize that and to give thought to possible remedies, which may or may not involve additional expenses. For those that do require funds, do the following:

Create a worksheet. Label it "Adverse Factors" at the top. In the left-hand column, list the items that will require money to address. If you have actual figures, include them. If you don't, assign someone to research them. Pencil in that person's initials next to the item along with the date that information should be ready. If any of these factors have the potential for preventing you from addressing a weak link or otherwise greatly reducing income, consider them wealth-generating expenses and record them on the Wealth-Generating Expenses worksheet you prepared earlier.

3. Are Current Enterprises "Profitable" and Have You Diversified Your Risk?

If the gross profit for any current enterprise is not proving to be as good as planned, you will need to determine why and make appropriate adjustments in the new plan or drop it in favor of a higher-yielding enterprise. On the other hand, if your cull ewes sold for considerably

more than you planned and expenses were right, what are the chances of that happening again in the next year? If the ewes brought top dollar because of your marketing efforts—something that *is* under your control—you'll be more confident in your predictions. If you lost an entire hay cutting to hail, and most of the next one to rain spoilage, what are the chances of that happening again in the next year? Though you can't control the weather, you do have some control over when to cut and bale the hay or whether to grow hay at all. Should you consider dropping that enterprise or researching insurance options?

Ideally, you want not only to maximize gross profit but also to diversify risk by not relying on one high-yielding enterprise to cover all your fixed costs. It would be unwise to commit all your resources—land, management time, and money—to hay production next year because its gross profit was exceptional this year.

You would have to decide how much of those resources you would commit and how much should be put into other enterprises to help diversify the risk:

- Do a gross profit analysis per 100 cows, acres, or whatever units are appropriate. Consider looking at what each enterprise contributes per dollar or per acre (or hectare) so you can better determine which enterprise to allocate more funds to in the coming year if you know you are limited by money or acreage (hectarage).
- Be prepared to drop any enterprise that is not providing a positive gross profit unless you have a special reason for running an enterprise that costs more than it brings in. It is the gross profits of all enterprises combined that must cover all the fixed costs of the business and provide the excess that becomes your profit.

This analysis is particularly important the first time you plan, since you may discover that one or more of your current enterprises is not providing sufficient income to cover fixed costs and is actually draining the business as a whole.

Create a worksheet. Label it "Gross Profit Analysis" at the top. List each of your current enterprises down the

Savory

Gross Profit Analysis

WORKSHEET

Date ___November 2018___

Planning Sheet
Column Reference _____ Worksheet No. _2_

	Jan	Feb	March	April	May	June	July	August	Sept	Oct	Nov	Dec	
PRESENT ENTERPRISES													
Corn	135,000												
Alfalfa	25,000												
Pigs	15,000	Plan to reduce expenses by free ranging with cattle.											
Chickens	5,000												
Cattle	20,000	Plan more income by selling pairs.											
NEW ENTERPRISE	BED AND BREAKFAST												
Income at 60% Occupancy	25,920												
Direct Expenses	6,480												
Gross Profit per annum	19,440												
Total	251,840												

Figure 1-12. The gross profit figures listed to the right of the first five enterprises are the gross profits expected by the end of the year. The notes to the right refer to modifications that will be made in the particular enterprise in the coming year to improve the gross profit. The new bed-and-breakfast enterprise under consideration shows the calculations used in estimating its gross profit. The total gross profit figure includes both the present and the new enterprises.

left side and record their gross profit. Note in the row to the right any modifications you plan to make in each enterprise in the coming year as a result of the gross profit analysis, as shown in figure 1-12. List any new enterprises you might be adding, as a result of either your market research or an earlier brainstorming session, and write in the gross profit figures you used in considering those enterprises.

With the gross profit analysis done, you have a good idea of which are your most "profitable" enterprises—the ones that contribute most to covering your fixed costs. And with all enterprises consolidated on the worksheet, you can see how profitable you are likely to be in the coming year. If there proves to be insufficient income to cover your fixed costs (which you will have some knowledge of from past experience), you are in trouble. If this is the case, now is a good time to revisit all enterprises and modify them or seek more lucrative ones (see box 1-4).

4. What Is the Weak Link in Each Enterprise and How Will You Address It?

Most likely, you have already given thought to what the weak link is in the chain of production for each enterprise—resource (energy) conversion, product conversion, or marketing (money conversion). Determine what actions you can take in the coming year to address it. Note that some or even all of your enterprises may have the same weak link, often the case with marketing, or that the actions taken to address a weak link in one enterprise may also address the weak link in others.

Occasionally, what starts out as the weakest link in a particular enterprise might be addressed quickly and with minimal expense early in the year, and another link then becomes weakest. Determine what action you will take to address that weak link as well.

Run each of the actions you could take to address the weak link through the following context checks, any

---

Box 1-4. Selecting Appropriate Enterprises

Before adding new enterprises, or eliminating old ones, bear in mind the following:

- *It takes time to develop skills and perfection in any enterprise.* The learning curve can be costly. As you change to the new, plan a solid overlap with the old where possible.
- *It is often easier to alter an existing product, or develop new uses for it, than to create something entirely new.* There may well be ways to "add value" to a soybean crop or a wool clip that increases the gross profit on that enterprise with little additional effort.
- *Possibilities may exist to collaborate or partner with other related businesses or organizations that produce complementary products or services.* You might combine with other producers to form a cooperative to better exploit a niche market, or contract to supply your sister-in-law's "whole foods" restaurant with organic lamb and new potatoes, smoked hams, and fresh corn.

- *Managerial effectiveness is diluted by the number of enterprises one manager is responsible for.* This is especially applicable to small operations with few employees. In taking on new enterprises, managers are often stretched too thin, which tends to destabilize all enterprises and the business as a whole. One way to overcome this problem is to contract the management of the new enterprise to someone else.
- *There is a direct relationship between management effectiveness and the distance to what is being managed.* This is the reasoning behind the old cliché that the finest fertilizer in the world is a farmer's footsteps, or the more current notion of managing by walking around. Obviously, the more frequent the contact with the enterprise, the greater your chances of spotting trouble early, and the more opportunities you have for finding ways to improve the enterprise.

---

of which could apply: cause and effect, weak link (social or biological), energy/money source and use, sustainability, and gut feel. Any expenses involved in implementing the actions that pass are considered wealth-generating expenses because they will strengthen the business overall. Thus these expenses should be given priority in your planning.

Create a worksheet. Label it "Weak Link Expenses." List all your enterprises in the left-hand column, leaving enough space between them to write in the actions that passed the checks. Determine the costs for implementing each action, and record the figures on the worksheet.

Putting the Plan on Paper

You've looked at any logjam that exists and any adverse factors, done a gross profit analysis of all your enterprises, and identified the weak link in each enterprise. You not only know in your heart what is right for you but you have checked to make sure the actions you plan to take are in line with your holistic context. Now you are ready to put figures to your dreams and craft the plan that will help you achieve your goals for the coming year.

Things to Keep in Mind

Few agricultural producers in the world enjoy the high prices for their products that Americans get for theirs. Even fewer worldwide enjoy lower prices for their purchased inputs or easier access to credit. Yet the failure rate of farmers and ranchers in the United States is a national scandal. The excuses are many, but the main reasons are these:

- Allowing production costs (despite low input prices) to rise to optimistically anticipated income
- Borrowing heavily against optimistically anticipated income
- Letting the promise of immediate profit mask the surety of damaging side effects
- Spending little time figuring out the reality on paper (most of those who do any methodical planning use conventional cash-flow techniques and economic concepts that do not work when the health of land, crops, people, and animals count as much as cash in the long run)

To counteract these normal habits of mind, do the following:

- Calculate your total gross income, then set aside up to fifty percent as your "profit." The remainder is all you have to allocate for the living and production expenses you are about to plan. This will discipline you to keep expenses below what you're actually likely to make. Although fifty percent may sound unreasonable, a number of ranchers have been able to allocate that amount and achieve it. You can, of course, allocate less than fifty percent, and many do, but allocate too small a percentage and you will expend too little effort in cutting unnecessary expenses.

- Heed this warning: Although you will deliberately seek to keep production costs from rising to anticipated income, your suppliers, salespeople, and consultants will not. They will still push strategies for "higher yields," "maximum gain," "heavy calves," and so forth instead of "greater profit." You can use their expertise, but make your position clear from the start: "Here are my limits. What can you do within them?" Almost anyone can push production up at increased cost; the challenge is to spread the difference between income and costs and thus push profit up.

- Keep your holistic context in mind at all times and check to ensure all the actions and tools you plan to use in running any enterprise are aligned with it. This will shift you away from the paper money trap toward solar dollars generated from the land itself, and away from simply living a life toward making a life based on what you and your family value most deeply.

- Work to a predetermined plan and schedule, and stay committed to others on the management team. This will force you to plan openly, thoroughly, and realistically.

Planning decisions that involve limits and trade-offs always take courage and discipline, but the process outlined here will help keep you and your planning team on target. Although you used the context checks to help ensure that your enterprises were sound, some of the tools and actions that govern each expense may need to be checked again as you work out the costs in detail.

The following steps will lead to a financial plan expressed in hard numbers. A worksheet showing how you arrived at your figures must back up all the numbers you enter on the income-and-expense spreadsheet. (Note: The instructions provided here assume you are creating your plan using the paper worksheets and spreadsheet. If you are using financial planning software you may need to make slight modifications. For instance, the paper forms use columns to reflect income and expenses on the spreadsheet, and rows to do the same on the worksheets. Some software programs do the opposite.)

Each income or expense column on the spreadsheet should be supported by at least one worksheet. Where numbers represent inventories such as fuel or feed as well as dollars, you must first plan consumption rates to show how you arrived at the dollar figures. All worksheets should be accessible, neat, and filed so you can later find them when you come to monitoring and controlling the plan to remain on track.

Step 1. Plan the Income and the Direct Expenses Associated with Each Enterprise

By now you should have a fairly good idea of what these figures should be. You will have used estimates in your preliminary planning; now you need to fine-tune them.

- **Create worksheets for each enterprise.** Fill out one income and one expense worksheet per enterprise, estimating the income and expenses and when they are expected to occur. Label each worksheet with the enterprise name, followed by the word "Income" or "Expense." For livestock breeding enterprises, use the Livestock Production Worksheet, which enables you to handle the complexity of births, deaths, and changes in age class during the year, and also records the costs of any animal purchases. The previous section has examples and directions. (Other costs, such as supplements, feed, veterinary expenses, and so on, would be recorded on an expense worksheet.)

Exploring Ways to Lower Your Costs

Profit is derived more from limiting expenses, and not allowing them to rise to anticipated income, than from any other factor in the business. Once you start putting your plan on paper, you'll almost certainly have to cut some expenses to achieve the profit you seek. That's why holding a brainstorming/brainwriting session, as described in "Mastering the Basics," every few years is a good idea. If you are going to be brainstorming new sources of income earlier in the year, use the same session and group of people to help you address ways to cut current expenses.

Concentrate on big, rather than small, expenses to get the most out of the session. Saving ten percent on your annual $2,000 phone bill doesn't begin to compare to saving ten percent on your $200,000 feed bill. Choose the top five expense categories or top ten individual expenses, or whatever is most helpful to you.

Start with the three-minute warm-up, make sure everyone understands the sort of expenses being considered, then brainstorm ideas as a group for ten minutes, and then individually for ten minutes. Or brainstorm each category of expense for three to five minutes as a group and then individually. To achieve the best results, you need to keep the session lively and fun. Make sure you run any of the ideas you consider implementing through the context checks.

- **Create inventory consumption worksheets.** Plan the consumption rate for any items you purchase in bulk that are a direct expense associated with an enterprise. This will tell you how much you have to buy and when. Record all figures in pencil because you are bound to alter them later when you have planned your profit and shifted your focus to the whole farm or ranch.

Step 2. Plan Miscellaneous Income

Think of any income sources not related to a specific enterprise, and then create a worksheet labeled "Miscellaneous Income." This could include mineral royalties, sales of junk or old vehicles, interest earned on a savings account, repayments on an employee loan, and so on.

Step 3. Categorize the Expenses

Your main task in this step is to think through each expense you will incur during the year and decide which category it will fall under:

- **Wealth-generating expenses** are those that will generate increased income this year, or soon after, and they could include those needed to address a logjam, an adverse factor, or the weak link in any enterprise. You identified these expenses in the preliminary planning and recorded them on the Wealth-Generating and Weak Link worksheets.

- **Inescapable expenses** are those you are legally or morally obligated to meet. They are fixed amounts that cannot be delayed or negotiated, making this category a minor one that may have no expenses allocated to it. Examples may include a debt obligation to a friend or family member, or a nonnegotiable land tax. Create an "Inescapables" worksheet if you have any inescapable expenses. List each item, and note payment amounts in the months they are due.

Creating a Depreciation Fund

An expense you might want to consider is an annual contribution to a depreciation fund, so that when an asset has lived out its lifespan, you have the money to replace it without having to borrow.

Create a worksheet listing the relevant assets, their potential lifespan, the cost of replacing them, how much you plan to contribute to the fund this year and in what months, and head it "Fixed Cost—Depreciation Fund." If you wish, you can actually "spend" this money by placing it in a tax-free, income-bearing investment until needed. Or, you may choose to invest in the growth of your own business by moving these amounts to wealth-generating expenses. By going through the exercise, you become aware of when machinery is likely to wear out and can plan when to replace it without having to borrow the funds to do it.

- **Maintenance expenses** are the remaining expenses associated with each enterprise, including those listed on the expense worksheets completed in step 1, and all fixed costs. They are essential to running the business and maintaining present income levels but will not generate additional income. Most expenses fall into this category: salaries and drawings, fuel, machinery maintenance, insurance, supplementary feed, and fertilizer, for example.

Step 4. Plan All Other Expenses

Other expenses include your "fixed costs," those costs you don't associate directly with an enterprise. Other expenses might also include those related to the "adverse factors" you identified in your preliminary planning that didn't qualify as wealth generating but were still important.

Some of these expenses can be grouped together on a single worksheet, labeled "Miscellaneous," but others will need to be grouped under headings on separate worksheets labeled accordingly. Items such as bulk fuel will require a separate inventory consumption worksheet to work out the estimated use of vehicles at different fuel consumption rates, times of likely bulk buying, and thus when money will be spent. It's generally a good idea to have a separate worksheet for "drawings" (and a separate column on the spreadsheet) if you are the owner of the business and don't pay yourself a salary.

Step 5. Transfer Income Figures to the Spreadsheet

Create an income column for each enterprise on the Annual Income and Expense Plan spreadsheet (illustrated in figure 1-13). The arrangement of columns bears some thought. If you intend to compute NMI, you might group together those income columns that fall under one person's supervision. Other ways to categorize the columns will no doubt occur to you. Create one last column for miscellaneous income.

Carry the income figures from each of the worksheets to the appropriate columns and months of the spreadsheet, and record them in the "Plan" row. Do it in pencil. A lot will change. If you're using a computer with a conventional spreadsheet program, make hand entries on a spreadsheet, as well, until you know the program has no bugs and you have confidence in it.

Total the income. Add the planned figures down each column and note the total in the "Planned Total" row at the bottom. Then add the planned figures across each "Plan" row and note the sum in a last column labeled "Total Income." The sum of the row totals and the sum of the column totals must match.

Step 6. Allocate Funds to the Logjam (If There Is One)

This is the one exception to the rule of planning profit before planning expenses because the business as a whole is at risk if a logjam remains unaddressed. If it takes money to address, label the first column on the expense side of your spreadsheet "Logjam." Transfer any expense figures from the Wealth-Generating Expenses worksheet associated with the logjam, and record them in the "Plan" row in the appropriate months.

Step 7. Plan the Profit

Your sole purpose in taking this step is to place a ceiling on how high your expenses can rise. To proceed, do the following:

- Reduce the gross total on the income spreadsheet by up to fifty percent, as shown in figure 1-14. Set that as your limit for the remaining expenses you are about to plan. This will discipline you to keep expenses below what you're actually likely to make.

- If you carry a big debt, the amount of money remaining after you cut the projected income and subtracted loan payments would be so small it would depress rather than challenge you. Since the point of cutting income by up to half is to encourage thrifty thinking, subtract the debt payment first and cut what's left, as shown in figure 1-15.

- Record the amount of profit you plan to set aside in the next column on your spreadsheet (or make it the first column if you had no logjam expenses). Label the column "Profit" and enter the figure in the "Plan" row under the months you plan to set it aside.

Annual Income and Expense Plan

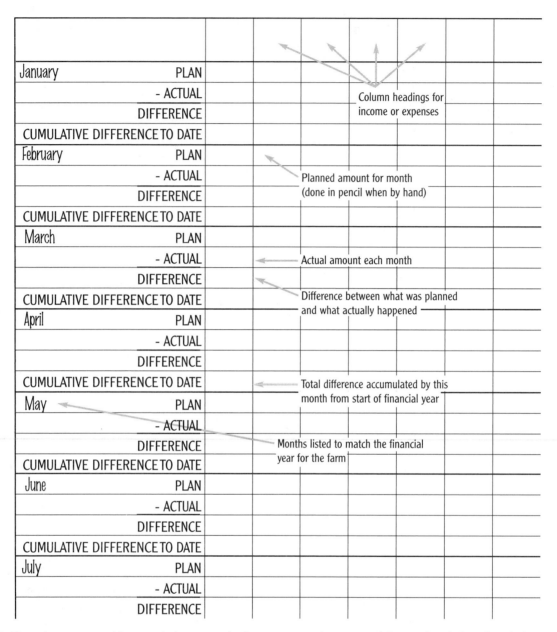

Figure 1-13. This is the master spreadsheet on which you enter the figures summarized on your worksheets and track them through the year. The actual sheet has 12 months vertically and 28 columns horizontally per page.

Step 8. Transfer the Inescapable Expenses (If Any) to the Spreadsheet

Allocate what you must, as these are by definition unavoidable and nonchangeable expenses; you do not need to agonize over them. Transfer the expenses listed on your "Inescapables" worksheet to the spreadsheet's next column, labeling it "Inescapables" and noting the figures in the "Plan" row in the appropriate months.

Step 9. Transfer the Remaining Expenses to the Spreadsheet

Transfer the wealth-generating expenses first. Divide these expenses into two groups: those that must have 100 percent of the money needed or they won't happen, and those that could work with a partial allocation. In the first type, allocate all that is needed. If a training program costs $1,000 and you allocate only $500, it won't happen, so allocate the full $1,000.

Planning Your Profit (No Debt)

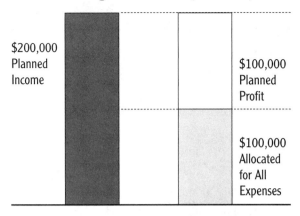

Figure 1-14. For some people, setting aside fifty percent of total income as profit, as this graph shows, provides a doable challenge. For others, fifty percent would be demoralizing—twenty percent or thirty percent would provide challenge enough.

Planning Your Profit (and Debt Servicing)

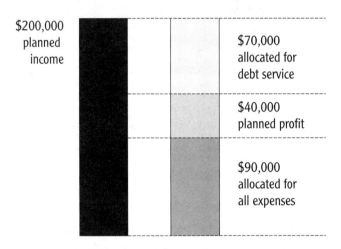

Figure 1-15. If a large portion of your income must go toward servicing debt, first subtract annual debt payments from the total income figure, and then determine how much of the remaining amount to set aside as profit.

For those wealth-generating expenses of the second type, make a good guess of the minimum funds you can allocate, with the full knowledge that the amount will increase before the planning is over. An example of this sort of expense would be fencing that is addressing a resource-conversion weak link. If you allocate the maximum dollars to this expense, you will run out of money before all maintenance expenses are covered.

Transfer any "Adverse Factor" wealth-generating expenses listed on the Wealth-Generating Expenses work-

sheet to a column labeled "Adverse Factors," and enter the figures on the "Plan" row in the appropriate months.

Transfer the weak link expenses associated with each enterprise into individual columns labeled "WL" plus the name of the enterprise. Because you will have other expenses associated with each enterprise, make sure you leave several blank columns following this one so you can keep enterprises grouped together. Count the remaining worksheets you have for each enterprise so you can determine the number of blank columns needed.

Transfer the figures on your enterprise expense worksheets to columns labeled with the headings used on each worksheet (see figs. 1-16 and 1-17). Don't forget to transfer monthly consumption figures in volume, number, or weight, rather than dollars, to their separate columns so you can monitor consumption throughout the year. This helps avoid the unpleasant surprise of having to bulk order fuel, for example, two months before planned. Record everything in pencil.

Transfer the expenses from the worksheets that list all your fixed costs. Make sure that each column carries the same name as the worksheet that backs it up. And don't forget to include columns for any inventory consumption worksheets. If you need additional columns start a second spreadsheet. Label the last expense column "Total Expenses."

Total the expenses. Check your figures. Total the columns and rows. The sum of the row totals and the sum of the column totals must be the same. Ferret out the mistakes and enter the grand total in the lower right corner of the spreadsheet. If you do this by calculator, use one that prints a tape, it makes finding errors vastly easier.

Step 10. Bring the Plan into Balance

Once all expenses are recorded and totaled on the spreadsheet, you will know whether your income covers them. If it does not, as is likely, your challenge now is to cut, paste, compromise, and fit everything into one holistically sound plan. In doing this you will inevitably modify some of the work you did in earlier steps, so don't take it personally when erasers get worn out on things you put a lot of time into. The work was not wasted. Let's say you figured out how to build thirty miles of fence but end up putting only twenty on the plan. You can't make that kind of decision at all without thinking through the bigger plan.

Expense Spreadsheet

		Gas Use	Bulk Gas $	Other Gas $
January	PLAN	71 g	1200	
	- ACTUAL			
	DIFFERENCE			
CUMULATIVE DIFFERENCE TO DATE				
February	PLAN	75 g		
	- ACTUAL			
	DIFFERENCE			
CUMULATIVE DIFFERENCE TO DATE				
March	PLAN	66 g		866
	- ACTUAL			
	DIFFERENCE			
CUMULATIVE DIFFERENCE TO DATE				
April	PLAN	74 g		
	- ACTUAL			
	DIFFERENCE			
CUMULATIVE DIFFERENCE TO DATE				
May	PLAN	63 g		
	- ACTUAL			
	DIFFERENCE			
CUMULATIVE DIFFERENCE TO DATE				
June	PLAN	63 g		
	- ACTUAL			
	DIFFERENCE			
CUMULATIVE DIFFERENCE TO DATE				
July	PLAN	63 g	1200	
	- ACTUAL			
	DIFFERENCE			
CUMULATIVE DIFFERENCE TO DATE				
August	PLAN	80 g		434
	- ACTUAL			
	DIFFERENCE			
CUMULATIVE DIFFERENCE TO DATE				
September	PLAN	80 g		
	- ACTUAL			
	DIFFERENCE			
CUMULATIVE DIFFERENCE TO DATE				
October	PLAN	92 g		
	- ACTUAL			
	DIFFERENCE			
CUMULATIVE DIFFERENCE TO DATE				
November	PLAN	63 g		
	- ACTUAL			
	DIFFERENCE			
CUMULATIVE DIFFERENCE TO DATE				
December	PLAN	63 g		
	- ACTUAL			
	DIFFERENCE			
CUMULATIVE DIFFERENCE TO DATE				
	PLAN TOTALS	853 g	2400	1300
	ACTUAL TOTALS			

Figure 1-16. The data from the worksheet in figure 1-17 would look like this when entered on the annual expense plan. Note that inventory consumption figures and actual purchases appear in separate columns, which enables you to monitor consumption so you can avoid unplanned-for expenses.

Start with the maintenance expenses. Cut out any that aren't absolutely essential to each enterprise or to running the business, or that can be put off for another year without damaging the enterprise or the business as a whole. You are challenging all maintenance expenses here, and nothing is sacred—even routine expenses don't necessarily have to continue.

Use the marginal reaction check. Apply any excess gains (from your maintenance expense cuts) to those wealth-generating expenses that received only minimal funds but needed every dollar they could get. Apply the most dollars where they give the highest return in terms of your business goals and your holistic context. Every dollar you can move from a maintenance expense to a wealth-generating expense without losing effectiveness will grow the business to a new level.

Use the marginal reaction check to compare the wealth-generating expenses across all enterprises. Identify those that provide the greatest return to the business as a whole, and be prepared to eliminate others. This suggestion is counterintuitive. You will be tempted to keep all wealth-generating expenses, but if you are serious about being profitable, don't.

Consider reducing your profit if you will be putting that money toward one or more wealth-generating expenses. In doing that, you make the decision to invest your profit in growing the business. Just make sure you're doing it intentionally and not just letting the expenses rise to the total income planned, as it is all too human to do. If you do intentionally make this choice, which is often wise, don't later complain about the lack of profit this year when you know you have lowered it purposely to increase profits in coming years.

Step 11. Determine Where to Invest Your Profit

When the plan is in balance and you have ensured that your income will cover all expenses, determine how best to invest the planned profit. If, in the previous step, you didn't consider reinvesting your profit in a wealth-generating expense, consider doing so now if you want to grow the business. You might also plan to invest some or all of your profit elsewhere to diversify your risk or to maintain some liquidity in case of emergencies. Bear in mind the tax consequences at the end of the year.

Worksheet Figures Transferred to Spreadsheet

Date 2018   **WORKSHEET**   Planning Sheet Column Reference Fuel   Worksheet No. 14

Seasonal Year	Jan	Feb	March	April	May	June	July	August	Sept	Oct	Nov	Dec	
Seasonal Year	← Feeding and Calving →				← Bulling →				Weaning	Sales	← Feeding →		**Item Totals**
Manager's Suburban	200m	250m	250m	250m	250m	250m	250m	250m	250m	400m	100m	100m	2,800 miles
Suburban Off Ranch			4,000m					2,000m					6,000 miles
Jim Jones Pickup	200m	200m	150m	50m	50m	50m	50m	50m	50m	50m	200m	200m	1,300 miles
Jones Motorbike			100m	100m	300m	300m	300m	300m	300m	300m			2,000 miles
Jake Moncho Pickup	350m	350m	250m	250m	60m	60m	60m	60m	60m	60m	350m	350m	2,260 miles
Moncho Motorbike			150m	150m	400m	400m	400m	400m	400m	400m			2,700 miles
John Smith Mechanic P.U.	100m	100m	100m	300m	300m	300m	300m	500m	500m	500m	100m	100m	3,200 miles
Pickups miles/ gals @ 12 mpg	850m/ 71g	900m/ 75g	750m/ 63g	850m/ 71g	660m/ 55g	660m/ 55g	660m/ 55g	860m/ 72g	860m/ 72g	1010m/ 84g	750m/ 63g	750m/ 63g	9,560m/799g
Purchases off ranch	4,000 miles estimated		$866		2,000 miles estimated			$434					$1,300
Motorbikes miles/ gals @ 90 mpg			250m/ 3g	250m/ 3g	700m/ 8g	700m/ 8g	700m/ 8g	700m/ 8g	700m/ 8g	700m/ 8g			4,700m/52 g
Monthly gas consumption	71g	75g	66g	74g	63g	63g	63g	80g	80g	92g	63g	63g	To plan sheet by month
500 gals bulk purchase gas		$1,200						$1,200 approximately 147 gallons on hand at year's end					$2,400
Monthly Totals	$1,200	0	$866	0	0	0	$1,200	$434	0	0	0	0	$3,700

Figure 1-17. All the expense and consumption figures have been transferred from this worksheet to the appropriate columns in the expense spreadsheet in figure 1-16.

Step 12. Check the Cash Flow

You have now laid out your proposed business year in its entirety. Take a good look at it. Even if the bottom line comes out positive, you don't know if you will have cash on hand when you need it or what your monthly credit needs will be.

To analyze your cash flow you will need to create two columns at the far right end of your spreadsheet, labeled "Monthly Surplus/(Deficit)" and "Bank Balance" (see table 1-2). If you know that you will need to borrow funds, whether as a line of credit or through overdraft facilities at your bank, or a note of any kind, skip this step and go to Step 13.

- *Monthly Surplus/(Deficit).* Calculate this for each month separately, and don't connect any one month to the next. Subtract the total expenses from the total income for the month, and record the result in the column for the month in the "Plan" row. If it is a negative figure—that is, more expenses than income—place parentheses around it.

- *Bank Balance.* Start with zero, or whatever cash you estimate or know will be on hand at the beginning of your financial year. Record this figure in the block just above the first "Plan" row. Next, add or subtract the monthly surplus or deficit you have planned for the first month. Enter the resulting planned bank balance, which will be the predicted bank balance at the end of the first month, on the "Plan" row of the "Bank Balance" column.

Successively calculate and record the figures for the remaining months. Make adjustments in any month in which income will be insufficient to cover expenses (the

Table 1-2. Cash flow columns

	Total Income for Month	Total Expense for Month	Monthly Surplus/ (Deficit)	Bank Balance $ *(Beginning Balance)*
Month 1 **Plan**			Income – Expense	Begin Bal. + Month's Result
Actual				
Difference				
Cum Diff YTD				

figures will appear in parentheses). You can do this most easily by rescheduling purchases, but some income items might be rescheduled to come in earlier.

Be careful about that, however. As a general rule, income tends to come in later than planned, and expenses tend to come in earlier than planned. Make sure that the calculations you use in making any adjustments are recorded on your worksheets before transferring the final figures to the spreadsheet. If you find you can't juggle figures sufficiently to avoid deficits over several months, you will have to borrow enough to get you through those months, either from overdraft facilities at your bank or by means of a short-term loan (covered in the next step).

Note: If you have made any adjustments, recalculate the total figures along the bottom of the sheet (for both income and expense columns) and across all "Plan" rows, and make sure the totals across and down are the same.

Step 13. Figure Debt Costs in Relation to Cash Flow (If Applicable)

If any borrowing is involved, you must work out the full cost of that borrowing by calculating the monthly interest. In addition, by manipulating the cash flow, you may be able to reduce the peak of indebtedness.

Do the monthly analysis of the debt on the spreadsheet, to the right of all the expense columns. A number of variations are possible. The number of columns you need for analysis will depend on your financing. If done through the bank by simple overdraft, you will need three columns. If done through the bank by a note or loan, you will need five columns. If you have more than one note or loan, you will need more than one group of columns. In all cases, you use essentially the same procedure outlined here.

Simple Overdraft

Use three columns headed "Monthly Surplus/(Deficit)," "Bank Balance," and "Bank Interest Owed" as shown in table 1-3.

In the "Monthly Surplus/(Deficit)" column, calculate the surplus or deficit for each month separately; do not connect any month to the next. Subtract the total expenses from the total income for the month and record the result in that month's "Plan" row. If it is a negative figure—that is, more expenses than income—place parentheses around the figure.

If you have no overdraft to start: Just below the "Bank Balance" heading, above the "Plan" row, record the

Table 1-3. Simple overdraft columns

	Total Income for Month	Total Expense for Month	Monthly Surplus/ (Deficit)	Bank Balance $	Bank Interest Owed
Month 1 **Plan**					
Actual					
Difference					
Cum Diff YTD					

amount of cash you estimate will be on hand to start the year. Then add or subtract from that amount the monthly surplus or deficit planned for the first month. Enter the resulting bank balance on the "Plan" row of the "Bank Balance" column. Then successively calculate and record the figures for the remaining months.

If you have an overdraft at the start or in subsequent months: From the "Bank Balance" column, take the figure of the last month of the previous year, or the first amount in overdraft (this should always be in parentheses to indicate a negative figure) and calculate the interest on the overdraft for that month by using the following formula: annual overdraft interest rate × planned overdraft divided by 1,200. For example, if the annual overdraft interest rate equals thirteen percent, and the predicted overdraft at the end of the first month equals ($20,342), then interest on the overdraft for the month would be calculated as follows:

$$13 \times \frac{\$20,342}{1,200} = \$220.37$$

Enter the interest calculated on the "Plan" row for the month in the "Bank Interest Owed" column (as shown in table 1-4).

To calculate the interest for the next month, start with the overdraft and interest predicted for the present month. Add that interest to the overdraft and then add the monthly deficit or subtract the monthly surplus for the next month. Enter the expected overdraft on the "Plan" row in the "Bank Balance" column, as in the following example:

Overdraft of month one = (20,342)

Add interest for month one of $220 = (20,562)

Subtract surplus in month two of $6,503
   = (14,059)

This amount becomes the planned overdraft or "Bank Balance" column amount. Then you can calculate and enter the interest expense of $152 for month two (as shown in table 1-5).

Assume that in month three the monthly income exceeds monthly expenditure by $15,000. The month's surplus is thus $15,000. That would give us the following calculation (as shown in table 1-6):

Overdraft end of month two = (14,059)

Add interest for month two of $152 = (14,211)

Subtract surplus in month three of 15,000
   = 789

The bank balance is now positive and is not entered in parentheses, nor is any interest calculated once the balance is positive. From here you would proceed with the calculations for the following months; but if you were to go into overdraft again, you would once more calculate and add interest charges.

Cooperatives and Similar Facilities
Some businesses use credit facilities offered by cooperatives or other lending sources that charge interest, which would give you the equivalent of overdraft facilities from more than one source. Although it's possible to predict the overdraft from each source separately, that complicates the arithmetic while not really improving the planning. We suggest you treat all similar credit facilities as one for cash management analysis, and that you use a single interest rate. Then use the same procedure you would use for a simple overdraft.

Loans and Notes
If you have a loan or a note, use an additional three columns to the right of the "Bank Balance" and "Bank Interest

Table 1-4. First month's interest owed

		Total Income for Month	Total Expense for Month	Monthly Surplus/ (Deficit)	Bank Balance (from zero dollars)	Bank Interest Owed
Month 1	Plan			(20,342)	(20,342)	220
	Actual					
	Difference					
	Cum Diff YTD					

Table 1-5. Calculating next month's interest

		Total Income for Month	Total Expense for Month	Monthly Surplus/ (Deficit)	Bank Balance (from zero dollars)	Bank Interest Owed
Month 1	Plan			(20,342)	(20,342)	220
	Actual					
	Difference					
	Cum Diff YTD					
Month 2	Plan			6,503	(14,059)	152
	Actual					
	Difference					
	Cum Diff YTD					

Table 1-6. When a deficit becomes a surplus

		Total Income for Month	Total Expense for Month	Monthly Surplus/ (Deficit)	Bank Balance (from zero dollars)	Bank Interest Owed
Month 1	Plan			(20,342)	(20,342)	220
	Actual					
	Difference					
	Cum Diff YTD					
Month 2	Plan			6,503	(14,059)	152
	Actual					
	Difference					
	Cum Diff YTD					
Month 3	Plan			15,000	789	
	Actual					
	Difference					
	Cum Diff YTD					

Owed" columns, headed "Loan Payments," "Loan Balance," and "Loan Interest Owed." Use the same procedure you would use for a simple overdraft, adding to or reducing the note as you wish in particular months to plan and calculate interest on the note balance. In the example shown in table 1-7, the opening debt is $100,000; the interest rate is twelve percent per annum, charged monthly; and the repayment is $2,000 per month. When an adjustment down is made to the principal amount of the loan in any month, that payment must be reflected in the "Bank Balance" column, because a reduction in loan liability will usually be offset by a decrease in funds available in the bank account.

Step 14. Assess the Plan

Your plan is not complete without first making two important checks:

1. Is the plan sound when projected forward? At this point your plan represents the actual cash flow into and out of the business, including the level of borrowing, if any, and when debt will peak during the year. Although the plan might look good now, you won't know if it really is good until you project it ahead several years to see whether it contains any distinct trend that leads to heavy borrowing. Doing this forward projection is relatively easy; it takes far less time if you have entered all

Table 1-7. When bank balance becomes positive

	Bank Interest Owed	Loan Payments	Loan Balance 100,000	Loan Interest Owed
Month 1 Plan		$2,000	$99,000	$1,000
Actual				
Difference				
Cum Diff YTD				
Month 2 Plan		$2,000	$97,990	$990
Actual				
Difference				
Cum Diff YTD				
Month 3 Plan		$2,000	$96,970	$980
Actual				
Difference				
Cum Diff YTD				

your figures onto computerized worksheets and spreadsheets.

To project the plan ahead, transfer the income and expense figures you have now to new worksheets. Then go through each item and make the changes that you know will apply in the following year. For instance, if you are running livestock and you included buying a scale and five new bulls in this year's plan, you know you can delete those expenses from next year's plan. On the income side you will need to rework the Livestock Production Worksheet so you can accurately assess the number of animals, hides, amount of wool, and so on for sale in the following year.

When you have made all the adjustments you can foresee and transferred them from the worksheets to the spreadsheet, simply look at the bottom line, the rate of payoff of any loans, and the peak of indebtedness during the year. If those look satisfactory, repeat this whole process for the year after. Continue to repeat this process for as many years ahead as you need to in order to get a clear idea of the trend of the business.

• **Is the plan sound from an overall business point of view?** Will the plan produce a profit or loss in real terms at the end of the planned period?

Remember that while you may have allowed for depreciation in real terms, you have not used government depreciation tables nor accounted for increases or decreases in net worth on the spreadsheets. Any items deliberately held back from sale, or unsold products held in storage at the end of the year, will represent an increase in wealth, although not yet reflected in cash.

To know how profitable the year is really planned to be, start by calculating the difference between the opening and closing (beginning and end of the year) values on all business assets, including unsold or held-back inventory. To the resulting figure add any cash excess planned and deduct the allowable depreciation on any assets. This final figure will give you your answer. Alternatively, create a beginning, or current, balance sheet listing total assets (current and fixed) and total liabilities (current and long term) and subtract the latter from the former to get the current owner's equity or net worth. And then create a projected ending balance sheet to see the change in your net worth—which should reflect a positive increase.

Don't panic. Think, think, think. Plan, plan, plan. Neither make, nor accept, any excuses. The word "can't" must not enter your head.

If you have any doubts, have your accountant review the plan and assist you; your spreadsheets and worksheets should contain all the information he or she will need. Knowing what the likely tax consequences of your plan are going to be will prove useful. It is for this reason that the planning should be completed at least a month prior to the end of the fiscal or tax year; it gives you the opportunity to push some expenses forward to the next planned year, or to arrange purchases before the end of your current year. In doing these things, you do not avoid taxation, but you can delay it to the benefit of your business.

If, after this final analysis, the plan looks good, then proceed. If not, replan right away. Replan until the results satisfy you. This replanning may require a great many adjustments and erasures of all those figures you wrote in pencil on the planning sheets (if you do everything by hand). Now, just when you need all the energy and creativity you can muster, you can easily fall into bitter arguments with your team members over how to cut and paste. Before that happens, consider the following: Many planning teams have found that extremely well articulated arguments often have nothing to do with the plan. They are trumped up mainly to avoid the agony of reworking the plan, which is something computers and financial planning software have helped overcome. You can just punch in the suggestions and see if they work.

Monitoring Your Plan to Ensure Profit

No plan ever goes exactly to plan. No planning is complete without monitoring, controlling deviations, and replanning when necessary. Before your new financial year begins, schedule monthly monitoring sessions and follow these four guidelines:

1. Establish the most convenient means to obtain, before the tenth of each month, the actual income, expense, and inventory consumption figures.

2. Enter your figures in ink on the second row (labeled "Actual") of the spreadsheet (see figure 1-18).

Monitoring Planned Versus Actual Figures

	Comm Cattle	Alfalfa Harvest	Alfalfa Income
January PLAN			
– ACTUAL			
DIFFERENCE			
CUMULATIVE DIFFERENCE TO DATE			
February PLAN			14400
– ACTUAL			18000
DIFFERENCE			3600
CUMULATIVE DIFFERENCE TO DATE			3600
March PLAN			
– ACTUAL			
DIFFERENCE			
CUMULATIVE DIFFERENCE TO DATE			
April PLAN			
– ACTUAL			
DIFFERENCE			
CUMULATIVE DIFFERENCE TO DATE			
May PLAN			
– ACTUAL			
DIFFERENCE			
CUMULATIVE DIFFERENCE TO DATE			
June PLAN		480 T	48000
– ACTUAL		400	41600
DIFFERENCE		⟨80⟩	⟨6400⟩
CUMULATIVE DIFFERENCE TO DATE		⟨80⟩	⟨2800⟩
July PLAN		480 T	24000
– ACTUAL		500	23400
DIFFERENCE		20	⟨600⟩
CUMULATIVE DIFFERENCE TO DATE		⟨60⟩	⟨3400⟩
August PLAN		480 T	
– ACTUAL		510	
DIFFERENCE		30	
CUMULATIVE DIFFERENCE TO DATE		⟨30⟩	
September PLAN			
– ACTUAL			
DIFFERENCE			
CUMULATIVE DIFFERENCE TO DATE			
October PLAN	9000		
– ACTUAL	9750		
DIFFERENCE	750		
CUMULATIVE DIFFERENCE TO DATE	750		
November PLAN	325156		
– ACTUAL	321025		
DIFFERENCE	⟨4131⟩		
CUMULATIVE DIFFERENCE TO DATE	⟨3381⟩		
December PLAN			13200
– ACTUAL			10600
DIFFERENCE			⟨2600⟩
CUMULATIVE DIFFERENCE TO DATE			⟨6000⟩
PLAN TOTALS	334156	1440 T	99600
ACTUAL TOTALS	330775	1410 T	93600

Figure 1-18. This plan shows the relationship between planned and actual income figures for cattle and planned and actual production and income for alfalfa.

3. For each month, compute the difference between the planned and the actual amounts. Enter the result *in ink* in the third row (labeled "Difference"). For all income and expense columns, enter figures that are adverse to plan in red ink and those that are on or better than plan in blue, so your overall position will be graphic.

4. After the first month, record the accumulated difference in the fourth row, again using red or blue ink. This is the sum of the differences in the preceding months. It will alert you when small differences ignored in your monthly control represent a serious drift away from your plan.

If you use the services of an outside accountant, you can't afford to wait for her to supply the figures you need each month. You know how much money you have spent and how much you have received and can record the amounts as the money comes in or as you commit to each expense. You don't need to nail down every cent. Monitoring quickly and on time matters more than perfect accuracy. If survival prospects look critical, then monitor actual figures *daily*, and project them to month's end (or even year's end) daily, to see where they stand relative to your plan.

One look at the planning sheet and its red and blue entries will show you the deviations from plan at a glance. Consider all the major adverse deviations in detail, column by column, going back to the original worksheets (which you have carefully filed), as necessary. At first the deviations may derive from inexperience, but as the years pass, you will get better. In any case, do something now!

If the income items are seriously adverse, they can be "controlled" only by cutting total expenses. This will require you to replan, perhaps from scratch. If the expense items are adverse, always control the item itself. You must never simply eyeball the numbers and balance a surplus in one column against a deficit in another. The general rule is to control deviations within each income or expense category rather than allow deviations between categories to balance out. Any deviation adverse to plan—fuel cost is a common one—must be controlled and brought back to plan. You cannot let a deviation pass on the excuse that a surplus for the month exists in some other category or column. You want to accumulate any surpluses you can, as they may be needed later in the year. If you don't keep every expense on target, you'll soon lose control of the whole plan.

It is crucial to apply the utmost energy and imagination to get things back on track. Mental attitude counts more than anything. Do not hurry.

Serious deviations from plan also constitute the greatest danger to management relations and the morale of your whole staff—another reason for careful monitoring. The temptation to assign blame quickly and to reprimand somebody will be great, but likely will ensure your ultimate failure.

The best solutions often lie in a flexible and cooperative sharing of time and resources commanded by several staff members, rather than merely turning up the heat on the one person in whose area the problem technically occurs. Anything you do now to exacerbate unhealthy competition, resentment, or turf battles among staff could prove fatal to your operation.

Make the problem a matter of shared concern and responsibility and invite a team approach to solving it. This is where ownership in the holistic context pays off in spades. Every crisis offers an opportunity for staff members to show their stuff. And seeking their input, rewarding creativity, and delegating authority build morale as surely as tyranny tears it down.

Generally—and certainly where a management team answers to a board or absentee owner—complete a monthly Control Sheet (shown in figure 1-19). This form states the column heading or number, the amount of deviation to date (in dollars, gallons, tons, etc.), the cause of the deviation, and the proposed action. Last, but most important, decide who is to act and note that person's initials in the "Act" column.

Consider devoting time in each monitoring session to look ahead to the coming months. Things you planned

Control Sheet

Name River Bend Ranch   Date June 2018   Sheet # 1

Plan Column #	Amount Adverse to Plan	Amount Adverse YTD	Cause of Deviation from Plan	Proposed Action to Return to Plan	ACT
I-5		$5000	Sales prices lower than expected.	–Plan reductions in all areas —	
				each person to prepare & come up with	All
				suggested reductions by next control meeting.	
				–Sell as pairs next year & market wider afield.	B.J.
E-16		$1500	Labor used earlier than planned.	–No action — will balance in August.	—
E-26		360 gals	Poor use of heavy transport.	–Reduce to 1 trip to town each month.	B.J.
			Too many unnecessary trips.	Keep purchase requests at office.	
E-37		$636	Farrier charges have risen	–Horses will not be shod. All hooves will be	
			as has transport cost.	rasped & checked weekly.	J.S.
				Send Jack on course in November.	
E-39		200 gals gas	Too much pickup use.	–Supplies will be bulked in small storage sheds at strategic sites on ranch.	
				–Use 4-wheelers — no pickup use in good	M.K.
				weather except emergency.	
E-42		$200	Several unexpected breakdowns	–Arrange brainstorming party Tuesday next	B.J.
			with machinery, pickups & tractor.	week at Bill's house 6 p.m. All staff present.	

Figure 1-19. The control sheet should be used for controlling major deviations from the plan, which become obvious as you tally the actual figures in each spreadsheet column each month. If minor deviations accumulate, which you will pick up in the "cumulative difference" row in each spreadsheet column, transfer the amount to the control sheet so it gets attention.

last November to have happen in June may need attention now to make sure they do happen as planned. The rains may not have come, and wool prices may have collapsed; the situation may be different from what you planned.

The plan will have a much greater chance of success if you go through it carefully each month, looking ahead to see what needs to happen, and noting those things in your time planner to make sure they do happen.

Summary

ALTHOUGH MONEY ITSELF IS NEITHER WEALTH nor happiness nor the fulfillment of dreams, it can be a measure of progress along the road to some of these things, and so you must plan. When farms and ranches representing the lives and dreams of generations go down to the mournful echo

of the auctioneer's hammer because no one saw the danger ahead, we talk of economics as an evil that grabs people by the throat in the dark. Yet you know that is not true. Just turn on the lights. Do the financial planning. The monster vanishes.

You can't manage in the dark. If you can't see all the pieces at once, even the notion of holism becomes absurd. Do the financial planning.

Allan Savory, who originated these procedures and has seen them return many an apparently hopeless case to the glory road of prosperity, often rails against a lack of self-discipline of well-endowed managers who don't plan. The people whose bones now bleach beside the road (more men than women) spent too much time drinking coffee and tinkering with machinery. Do the financial planning.

Most nonplanners probably don't lack self-discipline. When Sam Bingham doesn't plan, it's often because between him and those he cares for, be they family or associates, there is a no man's land of competition, disagreement, or ambiguity that they dare not enter. Nothing lights a fuse quicker than money.

"If my dad knew the size of our debt, he'd faint."

"Maybe, if we have enough left after the calves are sold, I'll get my daughter a car to take to college, or we can go visit my wife's parents, or lay that pipe from the spring."

If you wait till fall and the money is or isn't there, then you won't have to argue about it. Right? Or if there is an argument, the winners will spend the money and what happens then will be *their* fault. Either way, why stir things up now?

Because if you do, and you really can bring people together, create a holistic context, and eliminate the no man's land of silence, the planning won't require any self-discipline at all. Economics, which is no less a part of life than love, will be on your side. Switch on the lights and do the financial planning.

PART 2

HOLISTIC PLANNED GRAZING

HOLISTIC PLANNED GRAZING

Getting Animals to the Right Place, at the Right Time, with the Right Behavior

In the early days, people wishing to practice Holistic Management using livestock were usually advised to build an arrangement of fences, commonly but not always, in the pattern of a wagon wheel around a central water point. Although this is indeed an efficient way to handle livestock in many situations, the wagon wheel became such a strong symbol that it obscured the idea behind it. Like people who draw symbols on the doorpost to keep away lightning, many ranchers and researchers built wagon-wheel grazing cells in the belief that they had magic power to grow grass.

A legacy of the wagon-wheel dogma is the term *cell*, which Allan Savory first applied when planning an early project on a vast property that required dozens of fences radiating from numerous water points. On paper, the layout of the perimeter fences separating each water point looked like the top view of a honeycomb with its myriad cells. In that case, each water point and its herd were included in a separate grazing plan recorded on an individual planning chart. The term *cell* referred to the land area encompassed in each individual plan. But because the term *cell grazing* came into use and was linked to the myth about fence layout, *cell* was dropped in favor of the term we now use, *grazing unit*.

Planning for livestock grazing, not fencing layout, continues to define the grazing unit, which is any piece of land on which livestock moves are planned as a unit and recorded on a single grazing chart. You might have two or three grazing units on a ranch or one grazing unit encompassing an entire farm. Seasonally, if you change herd numbers or combinations of herds, so too may the area constituting a grazing unit for planning purposes change. You might, for instance, amalgamate two herds in two grazing units into one large herd running in one much larger grazing unit.

Art is sometimes defined as the representation of truth and beauty within artificial limits—a frame, the pages of a book, the range of an instrument. It is a daunting illustration of "progress" that not long ago whole continents were "grazing units" in which all interactions—between herds, soils, and plants—occurred without artificial limit, but today we measure grazing units in mere acres or hectares.

Planning for smaller land units demands considerable art, because such plans cannot be perfect, especially in the early stages of land development. Initially, because you are making decisions holistically and thus need to balance social, economic, and environmental factors, you may well have to settle for overgrazing some plants and higher-than-you-would-like levels of partial rest. You could, of course, avoid both these problems by increasing paddock numbers, but it wouldn't be economically sound to increase fencing before livestock numbers. You may, in fact, well find it impossible to apply all the tools to maximum effect until you can develop a sufficient amount of the infrastructure, such as fencing and water points, specified in your land plan. And the order and timing of that development will be determined by your financial planning through which you will coordinate the buildup of both livestock and infrastructure to ensure you remain profitable each step of the way. These constraints can prove discouraging enough to put some people off planning altogether. But holistic thinking and management mean embracing reality, not waiting for a perfect context. Only the experience of making the best plan for the season ahead can show you how to undertake long-range land planning to push back the limits and give yourself a more generous frame for your art in years to come.

Part 4, "Holistic Land Planning," addresses that broader challenge. Today, however, success lies in doing the best you can with what you have.

Mastering the Basics

CARRYING OUT THE FINANCIAL PLANS outlined in part 1 calls for a control of detail beyond anything generally seen in conventional budgeting. The main difference, however, is in the kind of detail you control. Although livestock may be critical to moving toward the goals for your business through profits derived from meat, milk, or fiber, raising livestock becomes a by-product of more primary purposes—creating a landscape and harvesting sunlight. This section deals with that challenge at the field level.

Since *Holistic Management: A Commonsense Revolution to Restore Our Environment*, third edition, explains the principles behind the grazing planning in depth, read it carefully, starting with chapter 41 ("Holistic Planned Grazing"). The bulk of this section provides a review of the major operational points and examples of how to apply them.

The Holistic Management framework puts all land management tools into four basic categories. Of these four, rest and living organisms (specifically grazing and animal impact) are the primary subjects of grazing planning (see *Holistic Management* chaps. 20–23). Four management guidelines govern most management decisions concerning these tools—population management (stocking rate), time, stock density, and herd effect (*Holistic Management* chaps. 37, 33, and 34, respectively). Division of a grazing unit into grazing areas (commonly called paddocks) is essential too. (Though these areas need not be fenced, we'll use the word *paddocks* for simplicity's sake.)

Why Plan Your Grazing?

The common excuses ranchers and farmers give for why they don't plan their grazing, or at least plan in any detail, are in fact the reasons why they must. If you knew exactly what weather to expect and had a perfect knowledge of the many other variables involved, then you might not need to plan—as long as you could keep all of that in your head all the time, which no human can do. When ranchers run into trouble with planned grazing the problem can often be traced back to their dropping the planning process altogether and simply rotating their mobs of animals through the paddocks. Or, in some cases, ranchers do plan but treat it as an event and not the continuous process that planning is—plan, monitor, control, and, if necessary, replan.

You plan your grazing so you can do the following:

- Create a landscape in line with your holistic context
- Achieve the profit you've planned this year
- Enhance animal performance and/or increase animal numbers
- Minimize overgrazing of plants
- Reduce or eliminate overrested plants and overrested soil surfaces
- In brittle environments, speed the biological decay (as opposed to chemical and physical breakdown) of dead and dying plant material

and in all environments to speed the cycling of mineral nutrients into the soil to promote the next season's growth

- Coordinate a variety of activities—for example, breeding, weaning, and so forth, and cropping, hunting, creation of firebreaks, personal schedules/vacations, and marketing
- Reduce labor and improve efficiency

Your planning will center on these questions:

- What landscape are you trying to create? (How did you describe it in your holistic context?)
- How much total forage will the land have to supply in the current planning period?
- How much forage will an average acre/hectare of land have to supply?
- How long will standing forage at the end of the growing season last in a nutritious state (including reserves for late springs, drought, fires, wildlife, and so on)?
- How long will animals spend in each paddock, and when will they return (the vital recovery period grazed plants require)?
- Where and when will you need to concentrate animals most to maintain healthy grassland, reduce weeds or woody vegetation, or prevent soil erosion?

Measuring Forage Utilization—in Days

Standard American practice measures forage use in "animal unit months" (AUMs). Months work well for computing grazing fees, but they prove clumsy for calculating whether a herd should move in two days or three. In this case, an "animal unit day"—shortened to "animal-day" (AD)—works much better to compute grazing taken in a single grazing period or over a season.

Animal-Days

An AD is the amount an animal eats in a day, and is thus first of all a measure of forage quantity, not animals or days. The concept of ADs helps you do the following:

- Estimate how much forage stock and wildlife will require from a given area of land

- Plan for dormant season and emergency reserves (assuming animals are outside year-round)
- Assess the impact of grazing on areas of different quality (or pressure on animals to graze poisonous plants)
- Determine a realistic stocking rate quickly and more accurately than other known techniques
- Cope with mixed species and classes of stock, when converted to standard animal units

In all your figuring, the AD will have the same character as bales, pounds, tons, cubic feet, or whatever unit you use for bulk measurement. Unlike a ton or a cubic yard, an AD of growing forage has no status at the Bureau of Standards.

Nevertheless, an AD, roughly speaking, is enough forage to fill an animal's stomach, as shown in figure 2-1. Your sense of what that means will grow quickly as you compute how many ADs your herd has taken and

Animal-Days of Grazing
Animal-days = Animals x Days

A cattle-day is the food a dry cow eats in one day.

A sheep-day is the food a grown sheep eats in one day.
About 5 sheep-days = 1 cattle-day

    1 cow x 100 days = 100 animal-days of forage
    10 cows x 10 days = 100 animal-days of forage
    4 cows x 15 days = 60 animal-days of forage

Figure 2-1. An animal-day is the amount an animal eats in a day, and thus is a measure of forage quantity, not animals or days.

observe the land before and after. Later, you will learn how to refine your estimates to take into account forage trampled to the ground, the presence of wild grazers, and other factors. For now, just think in terms of what one animal consumes in one day in bulk terms. Later we'll discuss how to cater for the physiological state of your animals at the time of grazing, that is, whether they are dry, pregnant and lactating, or needing to gain weight rapidly.

You will need to determine the ADs of forage you have both while planning and later when implementing your plan so you can better judge whether or not you have the forage your animals require. A cow eats more than a calf, and cattle eat more than sheep, goats, or donkeys, so if you're running more than one class of animal or cattle, sheep, and goats together you'll have to make some adjustments to provide one AD figure that enables you to assess the forage availability for all your animals. To do this, you can translate animal numbers into standard units that reflect the kind of animal and its size. Then add all the figures together to give you one AD figure, as shown in box 2-1.

Box 2-1. Conversion to Standard Animal Units

If you're running a breeding herd, or mixed-species herd, you will need to convert to standard animal units (SAUs) to better assess the volume of forage required to feed the herd. To do this, merely select one class of animal as your standard and then rate all the other species/classes to that standard animal.

The simplest way to do this is to use the following formula:

- A cow = 1 animal unit.
- A cow and her calf = 1.5 animal units.
- A weaned calf = 0.75 animal units.
- A bull = 2 animal units.

You can also rate a herd based on weights and percentages: a 1,000-pound (450-kilogram) cow is commonly used as the standard animal in the United States, but if your cows average closer to 800 pounds (324 kilograms), use that as your standard if it makes things easier. If a 1,000-pound cow is your standard animal, you can use the following formula:

- A 500-pound calf = 0.5, or 50 percent of an animal unit.
- A 750-pound steer = 0.75, or 75 percent of an animal unit.
- A 2,000-pound bull = 2, or 200 percent of an animal unit.

Use either approach when factoring in different species. The following formula is commonly used for small stock relative to cattle:

- 5 adult sheep/goats/pigs = 1 animal unit.
- 10 lambs/kids/piglets = 1 animal unit.

Here's how you would calculate animal days of grazing for a mixed herd of cows, bulls, calves, ewes, and lambs using the above formula:

- 100 cows = 100 animal units.
- 85 calves = 43 (85 × 0.5) animal units.
- 4 bulls = 8 (4 × 2) animal units.
- 200 ewes = 40 (200 ÷ 5) animal units.
- 300 lambs = 30 (300 ÷ 10) animal units.
- Total = 221 animal units.

If this herd spent 5 days in a paddock, it would consume 221 animal units × 5 days = 1,105 animal days of forage.

What these standard units don't do is factor in the physiological needs of the animals. A prize bull or lactating cow will have different forage requirements than a young steer. But don't assume it's just the bulk of forage that matters. A lactating cow generally won't eat much more than a dry cow, at least not enough to have a significant impact on planning. Her nutritional requirements are much greater, of course, but she will only eat until she is full. A dry cow will do the same. She'll just gain weight instead of produce milk. While simple field checks typically suffice, if you wish to use techniques that more closely determine an SAU, please refer to box 2-6, "Factoring in Physiological State and Wildlife Needs" later in this section.

Animal-Days per Acre (ADA) or Hectare (ADH)

Ideally, you'd like to know the exact amount of forage each acre or hectare could contribute to the ADs of forage your herd will take or require. The "animal-day per acre" (ADA) or hectare (ADH) is a practical alternative. Herds of different sizes may spend varying lengths of time in paddocks of different sizes, but you can still reckon how much forage the average acre or hectare in that paddock will supply, as shown in figure 2-2.

The example in figure 2-3 shows something of what this means in practice. In a winter's grazing, cattle have been through all the paddocks in a grazing unit three times. In one 500-acre paddock, 100 head spent three days, four days, and three days. The same 100 head spent one day, one day, and one day on a 100-acre piece of

Animal-Days Per Acre/Hectare of Grazing

Animal-days per acre (or hectare) = $\dfrac{\text{animals} \times \text{days}}{\text{acres (hectares) of land}}$

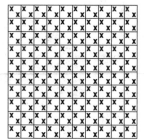

4 days
50 cows on 100 acres
(hectares)

Example: 50 animals spending 4 days in a 100-acre (or hectare) paddock will take $50 \times 4 \div 100 = 2$ ADA(H) from that paddock. An average acre (or hectare) will yield enough forage to feed 2 animals for a day.

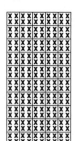

4 days
50 cows on 50 acres
(or hectares)

Example: 50 animals spending 4 days in a 50-acre (or hectare) paddock will take $50 \times 4 \div 50 = 4$ ADA(H). An average acre (or hectare) will feed 4 animals for a day.

Figure 2-2. Animal-days per acre or hectare tells how much forage the average acre or hectare in a paddock will supply.

Working Out the Animal-Days/Acre (or Hectare) the Land Yields

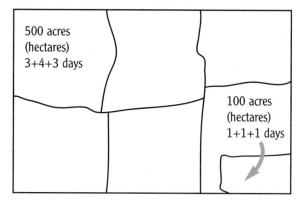

Land in the first yielded:

$$\frac{100 \text{ animals} \times 10 \text{ days}}{500 \text{ acres (hectares)}} = 2 \text{ ADA(H)}$$

Land in the second yielded:

$$\frac{(100 \text{ animals} \times 3 \text{ days}) + (100 \text{ animals} \times 1 \text{ day})}{100 \text{ acres (hectares)}} = 4 \text{ ADA(H)}$$

Figure 2-3. Animals have been through all the paddocks three times. In one 500-acre (or hectare) paddock, 100 head spent three, four, and three days. The same 100 head spent one day each on a 100-acre (or hectare) piece of bottomland, but on the last day they were joined by an extra 100 head gathered for shipping.

bottomland, but on the last day they were joined by an extra 100 head gathered for shipping. Both paddocks had acceptable amounts of litter and forage left after the last grazing. You want to know the relative production from these two paddocks.

Obviously if one paddock seems to respond differently (for better or worse) than others, you can compute the ADA(H) taken from it and change your plans in subsequent years.

Grazing, Overgrazing, and Growth Rates

Overgrazing is a term that applies to individual plants, not to grasslands, pastures, or paddocks. Simply stated, overgrazing occurs when an animal remains in an area for too long and grazes the same plant repeatedly as it tries to grow, or when the animal returns to a previously grazed area and regrazes a plant before it has recovered from the last grazing. The greatest damage to plants from overgrazing occurs when the plants are growing, especially when growing conditions are excellent. That's why scattering stock, or merely "turning them out" for a summer, as some American ranchers do on public lands, can do so much damage. This logic is counterintuitive, making it difficult to change ingrained practices.

A grazed perennial plant must utilize stored energy to grow new leaf before it can again reestablish the root or crown material from which energy was withdrawn to provide that growth. In general, the more leaf removed in a grazing, the more the energy that has to be drawn from the roots (or stem base or crown, depending on the species) so the plant can recover. Successive energy withdrawals, without intervening periods to reestablish root growth and/or energy reserves, can severely damage or destroy the plant. Overgrazing can be viewed as a repeated pruning of roots.

Overgrazing happens when animals linger too long among rapidly growing plants—or if they return too soon when growth is slow.

Where the thinking used to be that annual grass plants could not be overgrazed, recent research is changing that view. Annual grass roots and aboveground parts are in fact hindered in their development when grazed too frequently. Fortunately, overgrazing of annual grasses is generally avoided when you base recovery, and thus grazing periods, on the time it takes a perennial grass plant to recover from a severe bite. This holds true even when you are managing grasslands that appear to be made up entirely of annuals, such as those found in the degraded grasslands of Mediterranean-type climates.

Recovery Requirements

The recovery requirement varies with the site- and plant-specific growing conditions and the growth form (erect or runner) of the plant. Runner-type grasses tend to recover more rapidly than those with a bunched and upright form because a lesser percentage of leaf is removed in a single grazing of a given height.

A bunchgrass plant grazed down to three or four inches (seven to ten centimeters) will have had eighty to ninety percent of its leaf removed; but a runner grass

plant, which can cover many square yards or meters, will have had relatively little—perhaps ten to twenty percent—leaf removed if grazed to the same height. Thus runner grass plants under irrigation, or in weather conditions that produce rapid growth, can recover in as little as ten days.

At the other extreme, erect bunch grasses that tend to characterize very brittle grasslands that experience long periods of little or no growth, can take ninety days to a year or more to recover from a grazing. Climate and geography also influence recovery rates. Recovery is generally rapid in low-lying humid areas, for example, but much slower in high-elevation arid areas. Using times fairly typical for brittle environment grasslands worldwide of moderate to high rainfall, the diagrams in figures 2-4 and 2-5 show how this principle works.

When judging plant recovery times, it is better to build experience by actually observing the plants on your own place rather than to try using time periods suggested by others. Generally, bunched or runner grasses will have recovered enough to be grazed again when grazed plants look like ungrazed plants growing nearby. Some people have found it helpful to flag, or place a cage over, severely grazed plants and monitor their actual growth rate compared to the ungrazed plants.

There are also different degrees of recovery. For instance, a bunch grass grazed severely once in the growing season and once again in the nongrowing season will not be overgrazed. However, this plant will not develop as extensive a root system as it would have if only grazed once in the nongrowing season.

Recovery Time — Poor Growing Conditions

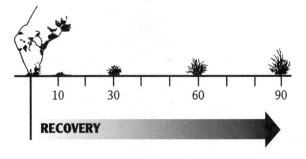

Figure 2-4. During poor growth conditions on rangeland, perennial bunch grasses may need ninety days or more to recover after being severely bitten.

Recovery Time — Good Growing Conditions

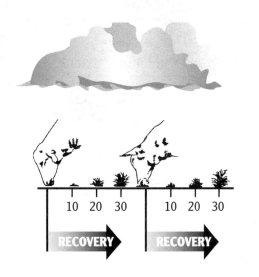

Figure 2-5. When proper moisture, temperature, and aeration combine, perennial bunch grasses may grow very fast and need less recovery time. Animals may bite them off after as little as thirty days.

Sometimes, or in some areas, you will be more concerned with high animal production and merely want to avoid overgrazing; in this case, you would opt for grazing it twice or more, accepting a lesser degree of recovery while not overgrazing.

In some paddocks you might be trying to achieve maximum root development to mitigate waterlogging or to break up deeper compaction; then you would opt for a single grazing and a greater degree of recovery. On the other hand, if much of the soil is capped and new plants are not establishing in some paddocks you may want to use shorter recovery periods so that the herd passes through those paddocks more often, breaking up more soil and laying down more litter to grow new plants. Bear such principles in mind as you manage to produce your future landscape.

If your holistic context requires a complex, stable grassland supporting many species, you'll adjust the timing to the most severely grazed grass plant and never to the average. The more any grassland is characterized by distinct plants easy to identify as single plants, the easier it is of course to judge a severely grazed plant. Severely grazed plants recover much more slowly than moderately grazed ones as the graph in figure 2-6 shows. You should

judge by the worst case (most severely grazed plant), even when the majority of plants have been grazed more lightly. This principle holds in any grassland or pasture situation.

The importance of judging by the most severely grazed plants, no matter how hard they are to identify, cannot be overstated. In a planted pasture situation especially, where severely grazed grass plants are easy to miss in the sward, farmers often do give inadequate recovery times. As a result, highly desirable nongrass plants, such as legumes, are killed out.

In principle, a recovery period between grazings as long as, or longer than, the entire growing season will enable most plants to fully recover from a severe grazing, regardless of daily growth rates. If you want to avoid any overgrazing during the grazing periods, then generally you must have enough paddocks to ensure that grazing periods are three days or less in any one paddock. In such a short time you will not overgraze plants and thus you do not need to worry whether growth is fast or slow. If your growing season was 180 days and you used that as your recovery period for running one herd through 100 paddocks, your grazing periods would average 1.8 days, and you should not overgraze grass plants, nor overbrowse woody or herbaceous plants.

Recovery and the Growth Curve

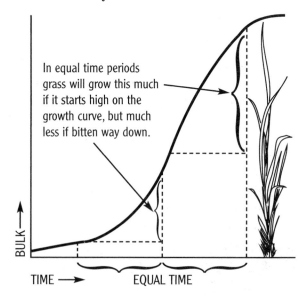

In equal time periods grass will grow this much if it starts high on the growth curve, but much less if bitten way down.

BULK →

TIME →   EQUAL TIME

Figure 2-6. Severely grazed plants recover much more slowly than moderately grazed ones.

When Recovery Rates Vary

In rangeland situations especially, a dilemma often arises over grasses that recover at different rates because of species characteristics, slope direction, or other factors. If you have only a few paddocks—and thus have to use long grazing periods in order to maintain recovery periods—you are likely to overgraze some plants because grazing periods are too long. Rapid moves favor the fast growers, but the herd returns too soon for the slow ones. Slow moves may expose fast growers to a second bite before they have recovered. Some plants will be weakened, and some may even die. But in brittle environments, in particular, partial rest (due to low animal impact) is far more damaging, in terms of the bare and capped soil that results, than the overgrazing of some plants.

To overcome the problem of varying recovery rates, as well as the high levels of partial rest, most ranchers try to build up to high paddock numbers as quickly as economic progress and planning allow, or turn to herding if the costs pencil out.

Grazing Periods and Recovery Periods

As chapter 33 in *Holistic Management* explains, *grazing and recovery periods are always linked.* Every time you shorten a grazing period in one paddock, you shorten the recovery period in all paddocks within that grazing unit and risk overgrazing many more plants. To minimize this risk, it's important to plan the recovery period first. Once you've determined what an adequate recovery period should be, that figure will dictate what the average grazing periods for each paddock should be. In a grazing unit of six equal paddocks, for example, a ninety-day recovery period dictates eighteen-day average grazing periods (figure 2-7).

During rapid growth, severely bitten plants might require only a thirty-day recovery period. Thus average grazing periods would fall to six days. The following is a rule of thumb, unless you have many paddocks per herd:

> Slow Growth = slow moves (and a longer recovery period).
>
> Rapid Growth = rapid moves (and a shorter recovery period).

Figuring Grazing Periods

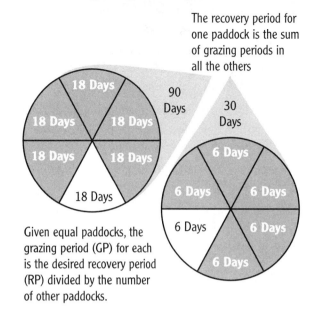

The recovery period for one paddock is the sum of grazing periods in all the others

Given equal paddocks, the grazing period (GP) for each is the desired recovery period (RP) divided by the number of other paddocks.

GP = RP ÷ (Total Paddocks - 1)

Figure 2-7. In a grazing unit of six equal paddocks, a ninety-day recovery period dictates eighteen-day average grazing periods.

When in doubt, use slow moves to minimize overgrazing. This rule runs counter to instinct. You naturally want to keep animals longer on dark green flushing grass and move on sooner when it's paler and obviously growing slowly—but as already mentioned, every time you leave a paddock a day early, it cuts a day off the recovery time for all paddocks. Thus, by the time you have moved through ten paddocks a day sooner than necessary, all paddocks have lost ten days of recovery time. Because this fact is so easily overlooked, all stock moves are planned on a chart, rather than in a notebook. The chart enables you to immediately see this loss of recovery time and avoid the overgrazing and loss of production that could result.

Time, Density, and Paddock Numbers

Timing and planning *on a chart*—these are the keys to Holistic Planned Grazing. As explained both here and in detail in *Holistic Management*, overgrazing occurs when animals take regrowing foliage before the plant has recovered from a previous bite. This happens when animals linger too long in one area or return to it too soon. Since, in all environments, plants grow at different rates, timing demands thought and vigilance even without such complications as calving and lambing, poisonous plant seasons, water scarcity, weather, and competing land uses.

In most situations—from small, tidy pastures to wide open, desertifying grasslands—land divisions are the instrument for striving toward what you need to achieve, whether or not you have fenced paddocks. These divisions can be demarcated by fences, natural barriers, or, for herders, blazed trees or natural features. In the future, technology may provide a practical form of virtual fencing.

Whether you're running a radial grazing unit with many fenced paddocks, leased land transected by ancient barbed wire, or unfenced grasslands, you will have to decide how long your animals will stay together in one place, how big that place will be, where they will move next, and when they will come back.

The Effect of Paddock Numbers on Timing

As the diagrams in figure 2-8 show, the more paddocks you have, the more recovery time per day of grazing each paddock gets. Increasing the number of paddocks doesn't change the ADA (ADH) yielded by the grazing unit as a whole. It just means that each acre or hectare gives up its share in a shorter time.

Once paddock numbers have built up to the point where the longest recovery period (slow-growth conditions) dictates grazing periods averaging no more than 3 days, you have the option of using a single recovery period—say 150 days, rather than 30 days for fast growth and 150 days for slow growth—which simplifies the arithmetic when you get down to planning. You can do this because even if the growth rate is fast there is little danger that plants will be overgrazed when exposed to animals for 3 days.

However, you may prefer to stick with using a range of recovery periods to better cater to the nutritional needs of your animals. Although you are not likely to overgraze plants over three days using that longer recovery period, you could get better animal performance during periods of rapid growth by dropping to a one-day grazing when growth is fast. Not only are the animals moving more rapidly onto fresh grazing, the plants they are selecting from

will be less fibrous. Depending on the number of days that growth was rapid before you again experienced slow growth and had to lengthen grazing periods, this would shorten all recovery periods to less than 150 days. This should still give adequate recovery time since all paddocks would likely have experienced rapid growth.

As the number of paddocks increases, the grazing periods become too short for overgrazing to occur, as mentioned, but other dangers increase, as figure 2-9 shows. Note that at thirty paddocks or so, slow moves during fast growth cease to cause overgrazing. However, if for some reason animals do return to any paddock before the plants in that paddock are fully recovered, you will experience great damage.

The reason is simply because the stock density will also be rising with decreasing paddock size, and so if any paddock is badly handled a great many plants will be overgrazed at the same time, leading not only to damaged plants but also, in a brittle environment, exposed soil. Once you attain high paddock numbers—up to 100 or more—the chances of returning before plants have recovered generally diminishes.

Moving too fast is a great temptation because a large number of animals crowded into a small area will deplete it quickly, and moving them a day early leaves a paddock looking much better and the animals happier. However, in a thirty-paddock grazing unit this cuts the recovery period in every paddock by a month, and the combination of both increased stocking rate and increased stock density may overgraze every single plant. The best way to avoid any of these dangers is to plan the grazing using a planning chart on which you are able to keep an eye on recovery periods for each individual paddock and the unit as a whole.

In general, most ranches are understocked as their managers begin to manage holistically and thus, in terms of Holistic Financial Planning, "product conversion" is the weak link. In these cases, the best marginal reaction per dollar comes from increasing animal numbers rather than paddock numbers. However, once the point is reached where the resource (energy) conversion link is weakest, the return on producing more paddocks—either fenced

More Paddocks Mean More Recovery Time/Day of Grazing

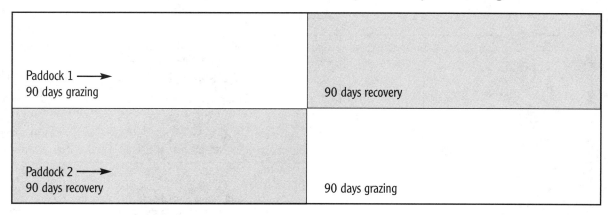

With 2 paddocks, each would get 90 days of grazing and 90 days of recovery in a growing season of 180 days.

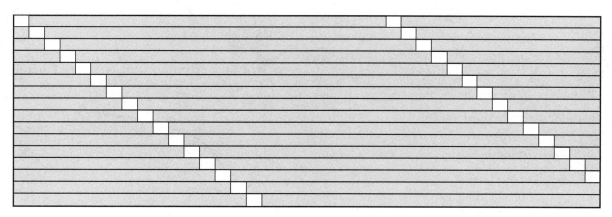

With 4 paddocks, each would get 30 days grazing followed by 90 days of recovery. Each of the first two paddocks grazed might then get a further 30 days of grazing followed by recovery until the 180-day growing season ends.

With 16 paddocks, each paddock gets 6 days of grazing followed by 90 days of recovery, with some (but not all) paddocks getting another grazing of 6 days until the 180-day growing season is over.

Figure 2-8. The more paddocks you have, the more recovery time per day of grazing each paddock gets.

Effects of Timing with Various Paddock Numbers

			Rapid Growth		Slow Growth	
Paddocks	Days/Grazing	Days/Recover	Grazing Period	Recovery Period	Grazing Period	Recovery Period
8	4	30	FAIR	GOOD	GOOD	DANGER too short
			(Few plants grazed at low stock density)			
8	13	90	DANGER regrowth regrazed	GOOD	GOOD	GOOD
				(More plants grazed as time in paddock increased)		
31	1	30	GOOD	GOOD	GOOD	EXTREME OVERGRAZING
				(More plants grazed at higher stock density)		
31	3	90	FAIR	GOOD	GOOD	GOOD
				(Even more plants grazed at higher stock density and longer time)		
91	1	90	GOOD	GOOD	GOOD	GOOD
				(Many plants grazed at very high stock density)		

Figure 2-9. As the number of paddocks increases, the grazing periods generally become too short for overgrazing to occur. However, moving too fast during slow growth cuts the recovery period by a month in a thirty-paddock grazing unit, and the animals may overgraze every plant.

or demarcated for herding—is amazingly high. Seldom, if ever, can the provision of additional water or any attempt to reduce heavy invasions of woody plants begin to compete dollar for dollar. It is because of such reasoning that while paddock numbers are low it generally pays to find an inexpensive way to confine animals to smaller units of land—such as strip grazing within the larger paddocks.

At some point water also becomes a limiting factor, but again be cautious in how you deal with it. Avoid rushing into costly water provision prematurely. Water, like fencing or livestock, should only have money spent on it when it alone will push production from sunlight to solar dollar further than any other measure.

Up to about thirty paddocks, each new paddock significantly shortens the average grazing period, as figure 2-10 shows. After that, the time gains (in terms of your ability to minimize overgrazing) taper off. However, as we will see, additional paddocks continue to increase stock density on a straight-line basis. As partial rest is commonly a greater problem than overgrazing on brittle environment grasslands, you will want to increase herd effect, if possible (or stock density through additional fencing), to help overcome it.

If you are running stock in a nonbrittle environment, the rule of thumb about maximum density and minimum time still applies, although for slightly different reasons. For instance, in tropical forests receiving 100-inch (2,540-millimeter) rainfall, farmers working with Allan Savory found that only by moving the animals very frequently could they get reasonable animal performance. These fast moves automatically came from high paddock numbers and higher density.

Paddock Number and Grazing Periods

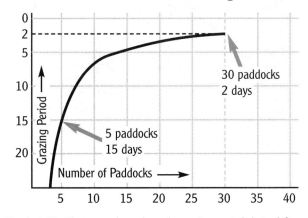

Figure 2-10. The curve shows how the grazing period derived from an average sixty-day recovery period shortens as the number of paddocks rises.

In any environment, increased paddocks will simultaneously improve the ratio of graze (and trample) to recovery because, although recovery periods don't change with increased paddock numbers, the grazing periods will always decrease. Here is one example:

If you have 10 paddocks and desire a 90-day recovery period in slow growth, average grazing periods would be 10 days (90 ÷ (10 – 1) = 10).

The graze/trample-to-recovery ratio would be 10:90.

If you have 20 paddocks and desire a 90-day recovery period in slow growth, average grazing periods would be 4.7 days (90 ÷ (20 – 1) = 4.7).

The graze/trample-to-recovery ratio would be much better at 4.7:90.

In desertifying, low-rainfall, brittle environments, the better this ratio, the more rapid the reversal of desertification tends to be, provided animal impact is adequate. And in nonbrittle environments the better this ratio the better animals tend to perform and the easier it is to slow the natural tendency of pastures to revert to forest. Aesthetic and legal considerations, type of terrain, cost of labor, herders, wildlife needs, and above all your holistic context, will control both the pace and the level of development. All this becomes easier the better you grasp the principles governing time, density, land, and animal performance.

The Effect of Paddock Numbers on Density

Stock density is the number of animals per unit of land (animals per acre or hectare) at a given moment. Since it involves both animals and acres or hectares, it influences the ADA(H) consumed from the paddocks within a grazing unit.

$$\text{Stock density} = \frac{animals}{acres\ or\ hectares}$$

The "acres" (and "hectares") in this equation do not refer to the total acres or hectares in the grazing unit, but the acres or hectares in a given paddock. Thus you could have high cattle numbers and low stock density or vice versa.

Because stock density includes the same animal and acre or hectare factors we use to compute ADA(H), it also represents how many ADA(H) the herd will take from the paddock every day they are in it. Every acre or hectare in a paddock stocked at ten cows per acre or hectare density will have ten ADA(H) removed every day the herd spends in the paddock.

At high density, therefore, the impact on an acre or hectare in one day becomes very great as illustrated in figure 2-11. If your planning is good, this will address many problems and speed land improvement. If your planning is sloppy or nonexistent, this will greatly increase the damage you can do to both land and animals.

The advantages of high stock density are as follows:

- Animals tend to graze a greater proportion of available plants and graze them more evenly, leaving fewer ungrazed.

- Distribution of grazing, dung, and urine becomes more even.

- Animals move more frequently onto fresh ground (because you have more paddocks), stimulating them and providing a more constant level of nutrition.

- Tighter plant communities tend to develop, providing more leaf and less fiber in many grass plants.

- Animal performance improves.

Note: If forage is growing and very rich, your animals' performance may actually drop due to lack of roughage. Or if recovery periods are very long, grasses may be excessively fibrous. In either case, you may need to use supplemental feeding as well as sound culling practices to assist in obtaining optimum individual animal performance in your herd. Good evidence also indicates that abrupt changes in diet often cause a drop in performance, as it takes several days for ruminal flora to adjust to the new regime. (Watch for circumstances where this would occur on moves from paddock to paddock.)

Paddock Numbers and Stock Density

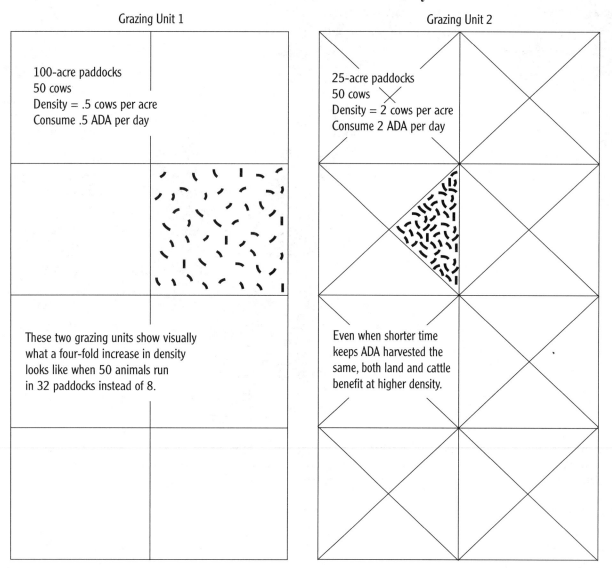

Grazing Unit 1

100-acre paddocks
50 cows
Density = .5 cows per acre
Consume .5 ADA per day

These two grazing units show visually
what a four-fold increase in density
looks like when 50 animals run
in 32 paddocks instead of 8.

Grazing Unit 2

25-acre paddocks
50 cows
Density = 2 cows per acre
Consume 2 ADA per day

Even when shorter time
keeps ADA harvested the
same, both land and cattle
benefit at higher density.

Figure 2-11. These two grazing units show visually what a four-fold increase in density looks like when fifty animals run in thirty-two paddocks instead of eight. Even when control of time keeps ADA (or ADH) harvested the same, both land and livestock benefit at higher density.

In your planning you'll probably consider increasing stock density in some areas before others and proceed by subdividing existing paddocks (see part 4 for details). When you do this you'll soon begin to notice the subtle relationship between stock density, paddock numbers, grazing periods, and recovery periods. As a general rule, smaller paddocks with higher stock density will begin to improve faster than large paddocks with lower stock density. The planning procedure in the next section ("Creating Your Plan") explains the mathematical rules. Box 2-2 demonstrates the principle.

Forage and Drought Reserves

In all environments ranchers and farmers will face adverse years when rainfall is short. In the more humid nonbrittle environments this shortfall may be slight, but relative to normal rainfall it could be serious enough to result in substantially less growth. Nonbrittle environments can also experience less growth in abnormally wet years due to drainage problems associated with a poor water cycle. In the more brittle environments, where rainfall is always erratic, dry years are more frequent and predictable.

Box 2-2. Effects of Splitting Paddocks

This grazing unit has six equal paddocks. A grazing period of twelve days in each will allow each paddock sixty days to recover. The herd will cover the whole grazing unit in seventy-two days.

Cutting paddock 1 in half to create paddock 7 will obviously double the stock density in these two paddocks. If, to keep the ADA(H) constant, you then graze the small paddocks only six days, you'll notice that they now get sixty-six days to recover instead of only sixty.

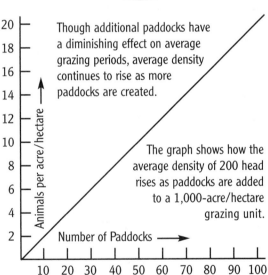

Though additional paddocks have a diminishing effect on average grazing periods, average density continues to rise as more paddocks are created.

The graph shows how the average density of 200 head rises as paddocks are added to a 1,000-acre/hectare grazing unit.

In practice, for the sake of convenience, we compute the grazing periods from a desired average recovery period rather than from a minimum, like sixty days in this example. The numbers don't look quite so neat, but the principle remains: when you subdivide paddocks, you reduce grazing periods and increase stock density and recovery periods in those paddocks.

Traditional practice calls for keeping some land in reserve to carry animals in time of drought, fire, or other catastrophe. It's reassuring to look over at tall grass waving in the reserved area and know it's there if you need it. On the other hand, owing to natural processes explained in the textbook, forage simply left standing for long periods of time loses nutritional value, and in brittle environments is more prone to oxidation and fire loss. In addition, overrested grassland in either brittle or nonbrittle environments promotes shifts to forbs and woody species.

Add up these problems and the traditional habit of setting aside land for drought reserve becomes an unwise practice. And if you don't use the reserves, when animals do graze those areas in the following growing season they will select new grass and leave the old to weather yet another year, again encouraging a shift to forbs and woody plants. Withdrawing some land from production reduces the graze/trample-to-recovery ratio, which puts heavier pressure on the remaining land, causing you to lose production over a larger area, and on animals that must now remain longer in paddocks.

Compared to the extra forage production possible when planned grazing encompasses all the land, withholding land for drought reserve typically results in a net forage loss and lower animal performance—a high cost to pay for a drought that may not occur. This practice of setting land aside for possible drought persists mostly as a defense against human nature. You know that in theory you need only one bank account, but putting something in a second account with a penalty for early withdrawal

SIDEBAR 2-1

A Note on Forage Digestibility

The more digestible the forage is, the more a cow can eat, assuming protein and mineral requirements are also met. The less fiber, or lignin, in the forage, the faster the rumen microbes can digest the forage and the faster the cow can empty and refill her rumen. The more lignin in the forage, the slower the rumen passage and the longer she stays full. That means she refills less; thus her total dry matter intake is less. This is why the better the forage the more the animals gain; not only can they digest more, they can consume more total dry matter.

helps suppress the temptation to dip into savings. A pasture withdrawn for drought reserve serves the same function, but in this instance it lowers the return earned on all money invested in all the bank accounts.

Since Holistic Planned Grazing offers a way to compute how many ADs a whole grazing unit can supply, you can budget a certain amount of days or months of grazing for reserve. Then you can handle the accounting separately on paper, just as the bank does, yet manage total capital, *on the land*, in the best interests of plants and animals at all times.

The example in box 2-3 illustrates the mathematics of reserving area versus reserving time and shows how plans that may be equal in terms of forage consumed result in quite different treatment of the land. Having extra paddocks available increases management flexibility as well as overall production of forage for a possible drought, and it enables you to get your animals through the drought period in better condition.

A simple technique allows you to estimate at any point how many ADs of forage you have on hand if growth should stop right now. Applied another way, the same technique will tell you how many ADA(H) the grazing unit will have to supply in order to carry your animals for a given time. Then you can evaluate sample areas against that figure and look far down the road as droughts

---

Box 2-3. Holding Reserves in Time or Area

Compare different reserve strategies for similar 1,000-acre (or hectare), ten-paddock grazing units. In terms of forage consumed, the plans seem equal.

A. Area-Reserve Grazing Unit

100 acres	100 acres	100 acres		
100 acres	100 acres	100 acres	100 acres	100 acres

B. Time-Reserve Grazing Unit

100 acres	100 acres	100 acres	100 acres	100 acres
100 acres	100 acres	100 acres	100 acres	100 acres

1,000 acres, 10 paddocks, 200 cows, 210-day growing season

Grazing Unit A grazes 800 acres and saves 200 acres for times when drought causes insufficient winter feed. In a good year, the 800 grazed acres should yield the following:

200 cows × 365 days ÷ 800 acres = 91.25 ADA

The reserved feed will be roughly 200 acres × 91.25 ADA = 18,250 ADs. The yield could be lower as it is only a nongrowing season yield when finally used.

Land in Grazing Unit B will yield the following:

200 cows × 365 days ÷ 1,000 acres = 73 ADA

Assuming grass grows equally in both units, each acre of the time-reserve unit will have roughly

91.25 – 73 = 18.25 AD of feed left.

Total reserves are again

1,000 acres × 18.25 ADA = 18,250 ADs

In a bad year, when reserves are necessary, all the numbers will be smaller, but the arithmetic will still show both plans being equal.

In terms of production, the time-reserve grazing unit will do much better:

1. Cattle in the area-reserve grazing unit take twenty-five percent more ADA from the land they graze, meaning more grass starts recovering from lower on the growth curve (see figure 2-6). Thus it regrows less in the same time than grass in Grazing Unit B.

2. Given an average sixty-three-day recovery period in each grazing unit, grazing periods (GPs) are two days shorter in Grazing Unit B.

   GP time = 63 ÷ 10 – 1 = 7 days (Grazing Unit B)
   GP time = 63 days ÷ 8 – 1 = 9 days (Grazing Unit A)

3. Cattle move to fresh grass twenty-five percent more often in Grazing Unit B, converting more of it into beef.

develop. Even more important, it allows you to plan for predictable nongrowing seasons, such as winter.

The calculations on the following page do not involve stock density at all, but rather stocking rate. *Density*, remember, refers to the concentration of animals in a paddock at a particular moment. Stocking rate is the number of acres or hectares you determine can support one animal for the time you expect a herd to remain in the grazing unit as a whole. This can be restated as the following equation:

Stocking rate = acres or hectares in whole grazing
unit ÷ animals in whole grazing unit.

For example, if you have 100 cows in a 10,000-acre unit continuously, that is a stocking rate of "1:100—year-round." A similar unit that carried 100 yearling steers only through the summer would have a stocking rate of "1:100—growing season only."

In planning, all stocking rates are handled similarly—except that when a grazing unit supports livestock for only part of a year, you'll have only the ADs required by wildlife to worry about the rest of the time. The examples in boxes 2-4 and 2-5 illustrate how to estimate the proper stocking rate for any piece of land. In addition, box 2-6, "Factoring in Physiological State and Wildlife Needs" is pertinent.

It takes some experience to decide what size square of land will support an animal for a day—and you must check several places on your land because there will be big differences between sites even on what superficially looks like a fairly uniform pasture or grassland. Records for past years (old grazing charts) will tell you what your land really does yield in ADA(H). Monitoring and fixed-point photos will also give you a feel for changes from year to year.

Roughly, however, if you can imagine yourself successfully filling a sack the size of an animal's stomach or making half a bale of hay while picking with only one hand, as a cow with one mouth and no hands would, you are in the ball park. If judging for animals that browse and graze, remember you will be looking at all the vegetation. Be aware that animals will be selecting their diets, taking leafy material of the season, some fibrous material, and

very little to no old oxidizing material, and they only have a limited time to do so since they have to ruminate and sleep. So don't simply consider the bulk of forage present, but also its quality. Eventually you'll find that just as a farmer who regularly weighs animals can estimate the weight of animals with remarkable accuracy, so too a farmer who regularly assesses the land in this manner becomes increasingly accurate.

In practice, when you go out on the land and lay out sample squares to see if your forage will last the winter, if all is well, there should be no doubt and the answer is yes. If you find yourself fudging a bit on whether a sample will feed an animal for a day, or dragging your feet on testing areas that you suspect fall below the average, you're probably in danger. As a general rule of thumb, any time you hesitate call it a no.

If, after sampling a good number of squares, you find that on average they would not provide the required feed, your next step is to calculate how many animals you can run. What size square will carry an animal for a day? This gets pretty tricky. Even though the procedure outlined here provides the best information you are likely to get, it's still very crude. A few yards can make a huge difference in your calculation of total reserves, so be prepared for some real soul-searching, and don't be afraid to get multiple opinions and to always err on the conservative side.

If your paddock arrangement allows you to do it in a relatively short grazing period and it would not disrupt your plan too much, you might plan to take all the nongrowing season ADs planned for one paddock in one grazing toward the beginning of the season. That will give you a very precise reading on how well you estimated for that paddock, but you can use that evidence to check your estimates in others.

Although we stress the difficulties and dangers, experience has shown over the last forty years that the worst mistakes managers have made when planning their grazing are nothing like the serious mistakes routinely made by managers who don't do the planning.

Determining Correct Stocking Rates

The technique of measuring forage available by checking the area needed to feed one animal for one day can

Box 2-4. Figuring Approximate Stocking Rate (Acres)

Imagine a sheep ranch in very poor country with a stocking rate of about 1:20. That means twenty acres for every sheep. To keep the numbers simple, assume you're keeping no lambs through a winter, which lasts from mid-October to mid-April (180 days). Thus every twenty acres has to supply 180 sheep days, or animal days (ADs), of winter feed:

180 AD ÷ 20 acres = 9 sheep-days per acre (ADA)

So, one-ninth of an acre must be able to feed one sheep for one day.

Now look at your land after first frost and see if one sheep could eat for one day on a selection of one-ninth-acre samples.

1 acre = 4,840 square yards.

⅑ acre = 4,840 ÷ 9 = 538 square yards.

Pushing the square-root button on a pocket calculator will quickly tell you that one side of a square covering 538 square yards is 23 yards. You can then step off several random squares (see fig. 2-12), and in each case ask: *Could this square feed one sheep for a day?* (It helps greatly if four people do the pacing and can stand at each corner while you judge whether the area *could* feed one sheep for a day).

You can also work the same problem the other way to find out what stocking rate your land *can* carry through the winter. After a very dry summer, suppose you find that a square of 23 yards to a side will not feed one sheep for one day. Suppose you find that the square has to be at least 35 yards on each side. From this you can determine how much stock you must sell to get through the winter (or how many sheep days of feed you must buy):

35 yards × 35 yards = 1,225 square yards needed
    to feed one sheep for one day.

Therefore, to find out how many sheep one acre would support for a day, you divide 4,840 (square yards in an acre) by 1,225 (area required for one sheep) to give you the answer—3.95, or four sheep. (Note: You need to subtract obvious areas such as roads or steep hillsides that will never be grazed from your total land area. This ranch might have a total land area of 42,000 acres, but only 40,000 would actually be grazed and thus considered in your estimates.)

If 160,000 ADs feed only 889 sheep for 180 days (160,000 ADs ÷ 180 days = 889 sheep), and if you normally run 2,000 sheep, you can expect them to run out of forage in about 80 days (160,000 ADs ÷ 2,000 sheep = 80 days).

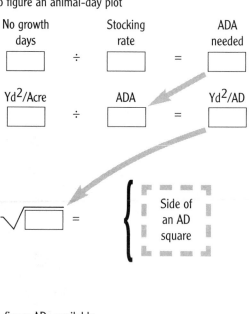

To figure an animal-day plot

| No growth days | ÷ | Stocking rate | = | ADA needed |

| Yd²/Acre | ÷ | ADA | = | Yd²/AD |

√ ☐ = { Side of an AD square

to figure ADs available

| Side of an AD square | ² | = | Yd²/AD |

| Yd²/Acre | ÷ | Yd²/AD | = | ADA available |

| ADA | x | Total acres | = | Total ADs |

| ADs | ÷ | No-growth days | = | Herd sustainable |

Box 2-5. Figuring Approximate Stocking Rate (Hectares)

Imagine a sheep ranch in very poor country with a stocking rate of about 1:5. That means five hectares for every sheep. To keep the numbers simple, assume you're keeping no lambs through a winter, which lasts from mid-April to mid-October (180 days). Thus every five hectares has to supply 180 sheep days, or animal days (ADs), of winter feed:

180 ADs ÷ 5 hectares = 36 ADs per hectare (ADH).

So $\frac{1}{36}$ hectare must be able to feed one sheep for one day.

Now look at your land after first frost and see if one sheep could eat for one day on a selection of $\frac{1}{36}$-hectare samples.

1 hectare = 10,000 square meters.

$\frac{1}{36}$ hectare = 10,000 ÷ 36 = 278 square meters.

Pushing the square-root button on a pocket calculator will quickly tell you that one side of a square covering 278 square meters is 17 meters. You can then step off several random squares (see fig. 2-12) and in each case ask: *Could this square feed one sheep for a day?* (It helps greatly if four people do the pacing and can stand at each corner while you judge whether the area *could* feed one sheep for a day).

You can also work the same problem the other way to find out what stocking rate your land *can* carry through the winter. After a very dry summer, suppose you find that a square of 17 meters to a side will not feed one sheep for one day. Suppose you find that the square has to be at least 25 meters on each side. From this you can determine how much stock you must sell to get through the winter (or how many sheep days of feed you must buy):

25 meters × 25 meters = 625 square meters needed to feed one sheep for one day.

Therefore, to find out how many sheep one hectare would support for a day, you divide 10,000 (square meters in a hectare) by 625 (area required for one sheep) to give you the answer—16 sheep.

A 10,000-hectare ranch supplies 160,000 ADs (16 ADH × 10,000 hectares = 160,000 ADs). (Note: You need to subtract obvious areas such as roads or steep hillsides that will never be grazed from your total land area. This ranch might have a total land area of 10,100 hectares, but only 10,000 would actually be grazed and thus considered in your estimates.)

If 160,000 ADs feed only 889 sheep for 180 days (160,000 ADs ÷ 180 days = 889 sheep), and you normally run 2,000 sheep, you can expect them to run out of forage in about 80 days (160,000 ADs ÷ 2,000 sheep = 80 days).

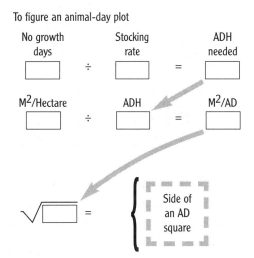

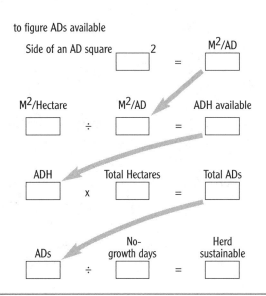

Box 2-6. Factoring in Physiological State and Wildlife Needs

Factoring in the physiological state of the animals is important if they are heavily lactating cows coming to the bull, for instance, or young stock on which you want to see high daily weight gains. In earlier days, we used to convert to standard animal units, or SAUs, using sophisticated tables that factored in the class, weight, physiological condition, and desired weight gains of the animals.

In practice, however, it has often proved just as effective to convert to SAUs based on weight or size only, as explained in box 2-1, "Conversion to Standard Animal Units," and to then factor in physiological condition or needs when assessing forage samples in the field. All you do then is rephrase the question you ask about feeding an animal for a day. When you look over each sample square of forage, pose the question as follows:

- If all your animals are dry, pregnant cows and you have no wildlife populations also dependent on the same forage, simply ask, Could this square feed one animal today?
- If the bulk of your cows are lactating and thus require a higher plane of nutrition, or if you have steers needing to gain weight, ask, Could this square feed one animal very comfortably today?
- If you also have a significant population of deer, antelope, or other wildlife, ask, Could this square feed one animal comfortably today and leave enough forage for the wildlife?

- If you also want to ensure abundant cover remains for game birds or other wildlife, ask, Will this area feed one animal and leave plenty of cover?
- If you are trying to cover soil as rapidly as possible, you might ask, Will this feed one animal today and leave plenty for litter cover?

Vary the question depending on the needs of your stock, your wildlife, or your land. When asking the questions, particularly in the case of lactating animals or those from which you are expecting high weight gains, it helps to "think like" a cow/goat/sheep or whatever the animal you are judging for as you look at that sample square.

To do this, imagine you have a sack the size of the animal's stomach tied around your neck that you have to fill in a limited time with an exact percentage of fresh, leafy material, and an exact percentage of fibrous material, and you only have one hand to do the filling. If, in looking at the plants on your sample area, you feel this would be difficult due to the sparseness of the forage, or an excessive amount of old, oxidizing material or fibrous stems, then it would also be difficult for a cow, sheep, or goat.

If you're running mixed herds of lactating mothers you hope to breed, plus young stock you want to gain, as well as culls, what do you do? Or if you run a mixed herd of sheep, goats, cattle, or whatever, and they are in differing stages of lactation or breeding, you will be called upon to apply that rarest of all human traits—common sense. If in doubt, err on the side of caution.

---

also help you during the growing season to determine if your stocking rate is reasonable. Start by sampling the longest recovered paddock that animals are soon due to graze. Compute the size of the square necessary to feed one animal for one day during the shortest grazing period allowable for that paddock. Because you would use that size square only during ideal growing conditions, you are certainly overstocked if that size square does not pass the test. See figure 2-12.

For example, suppose you plan 800 cows for a 600-acre paddock on three-day (fast growth) to nine-day (slow growth) grazings:

800 cows × 3 days ÷ 600 acres = 4 ADA.

4,840 square yards per acre ÷ 4 ADA = 1,210 square yards per AD.

Square root of 1,210 square yards = 35 yards (one side of an AD square).

Estimating the Area Required to Feed One Cow One Day

Figure 2-12. Two people have paced out a sample square they believe could feed one animal for one day. Markers aren't necessary when you have four people doing the pacing.

Or, for an example in hectares, suppose you plan 800 cows for a 200-hectare paddock on three-day to nine-day grazings:

800 cows × 3 days ÷ 200 hectares = 12 ADH.

10,000 square meters per hectare ÷ 12 ADH = 833 square meters per AD.

Square root of 833 square meters = 29 meters (one side of an AD square).

If this size square would feed one animal for a day, compute the smaller square that would support one animal for one day, given the longest grazing period in that paddock (nine days in this case) (see fig. 2-13). If that doesn't pass, you're probably gambling on rapid growth. Monitor carefully.

If animals seriously deplete forage or start to eat litter before they're scheduled to move, suspect overstocking but check other factors:

- You might have overestimated a paddock's capacity. Taking more ADA(H) from a better paddock might relieve the pressure.

- Low stock density can produce forage too rank to eat. Mowing, burning, or herd effect can solve that problem, but plan for higher density as soon as you can.

In year-round grazing situations, overstocking usually takes its toll only when animals exhaust forage during the nongrowing season. Thus, if overgrazing in the growing season was minimized, overstocking hurts animals, wild and domestic, more than it affects plants or the land. (In the unplanned grazing of old, overstocking hurt the land before the animals because plants were being overgrazed, and the land was deteriorating as a result. Now, under Holistic Planned Grazing, overstocking causes the animals to fall off in condition before the land because the

Longer Grazing Periods Mean Smaller Sample Areas

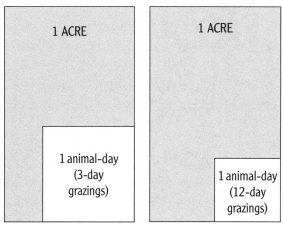

Figure 2-13. The longer the grazing period, the smaller the square that must provide each animal-day of forage.

animals don't have enough to eat, even though plants are not overgrazed.) Consider the following:

- As reserves dwindle, grazing animals will eat litter, reducing the soil cover so essential to the healthy functioning of all four ecosystem processes.
- Scarcity of plants tall enough to penetrate snow harms species such as pronghorn that do not forage through snow. (Plants that penetrate snow cause it to melt faster and may actually open up patches of grass. They also hold blowing snow, returning more moisture to the soil).
- Certain animals will shift exclusively to browse, creating "browselines" below which younger, shorter animals can find nothing.

These problems appear most frequently in yearling operations that do not plan to keep livestock through the dormant season and do not budget a winter reserve for wildlife. Even so, year-round grazing plans must recognize the danger too.

If you do intend to manage for any significant wildlife population at all (remember, diversity is almost always an asset), careful planning of forage will affect game populations more than any single factor, including hunting. Forage and cover governs how many wild animals survive the nongrowing season and where they spend it. An elk herd breaking fences to get to a haystack can cost plenty.

Use your imagination to develop a way to train yourself to judge more accurately by subjecting one paddock to the grazing pressure you figure all your land would have to stand, on average, during the winter, but then fence out some sample squares representing what you estimate would have to feed one animal for one day at that level of use (fig. 2-14). At the end of the test period you then can look at the grazed land and the ungrazed squares and discuss how close you have come. If the ground that was grazed still contains acceptable litter and forage, you can expect that similar use through the winter would benefit the land, and cattle performance would not suffer. Then you could use the fenced-off patch as a proper example of a square that could indeed feed one animal for one day at that level of use.

The Critical Slow- or Nongrowing Season

An underlying principle of Holistic Management holds that naturally functioning wild herds do not destroy land. Animals may die off occasionally because of drought, late

A Cow-Day Enclosure

Figure 2-14. To check estimates of nongrowing-season forage reserves, one option is to use a cow-day enclosure. Several enclosures can be placed in a paddock grazed at the grazing pressure you figure all your land would have to stand, on average, during the winter; then you can compare the grazed land and the ungrazed enclosures to see how good your estimates have been.

springs, and other climatic surprises, but almost by definition these setbacks occur when plants are not growing and the danger of overgrazing is minimal. Thus wild population levels generally reflect the amount of forage and cover available in the nongrowing season.

For domestic stock also, it is the amount and quality of standing forage available at the onset of the nongrowing season (or in some less brittle environments the slow-growing cold season) that largely determines your stocking rate. No other factor tells as much about the carrying capacity of your land.

If, as in a yearling operation, your land doesn't have to carry a herd through the nongrowing period, regulate your numbers to leave enough ADs of forage for wildlife and some reserve against a late return of growing conditions. But when you carry animals year-round your numbers will have to reflect what you can carry through the nongrowing period plus a reserve for wildlife and drought.

Only in the most extreme cases of overstocking combined with bad grazing planning do animals run out of feed during a growing season. In brittle environments, far more typically a dry summer does not leave enough standing forage to last through the winter. In the Northern Hemisphere, if the supply runs out in February and you don't expect significant growth until the end of March, you're in trouble. Good planning will cater for this by treating every year as though it is going to be a dry year—as fifty percent or more are likely to be in low rainfall, brittle environments.

Droughts are often just as predictable as dry years. In northeastern Arizona, for example, ranchers can expect some moisture in March or April, but if mid-May arrives dry, they had better figure out what their stock will eat and drink until August. The odds get very long in June and July. Again, good grazing planning will help you avoid such a calamity.

The condition of the land will tell you much more about the likely frequency of droughts than sophisticated weather models ever can, simply because droughts are largely the result of noneffective rainfall, rather than total rainfall. You can judge the degree of your own rainfall effectiveness simply by looking down at your land as you walk across it and noting the amount of bare soil between the plants. The higher the percentage of bare soil, the less effective your rainfall, and the more likely you are to experience droughts.

If You Run Out of Feed . . .

Whenever you think you may run out of feed during a time of minimal or no growth, keep these points in mind:

- Before cutting stock, do field checks to see how many ADs of forage you actually have. Use the "fence-out" technique described in figure 2-14 if it will make you more confident in your assessment.

- Consider changing plans to enhance any growth that plants do achieve and amplify the rebound when good conditions return. This generally means combining stock into fewer, larger herds. In fact, seriously consider combining all herds into one to extend recovery periods and keep animals moving as frequently as possible.

SIDEBAR 2-2

The Van Vuuren Case

In the 1960s in Zimbabwe, when Allan Savory was pioneering his grazing planning ideas, he came to the assistance of two brothers, Ben and Boet van Vuuren, who had four ranches bordering one of his own. The rains had failed completely, coinciding with a massive invasion of armyworms that had devoured everything in their path but coarse sticks. Ranchers in the area could not destock, as they had nowhere to send their cattle. Many had resorted to shooting their animals.

Desperate, the Van Vuurens asked Allan for help. Lacking any experience in a situation this bad, he fell back on an observation of French pasture specialist Andre Voisin that animals performed better on constant moves to fresh ground. He suggested the Van Vuurens amalgamate the eight herds on their four ranches into one herd of over 1,000, and plan daily or half-daily moves through all the paddocks. The four ranches, in effect, became one large grazing unit.

Over the next six months or so the animals subsisted on twigs and leaf fall, but did appear to receive some stimulation with the daily moves to fresh, unfouled ground. Although they were desperately thin by the time the rains broke, every animal made it through. Nearby ranchers who had half the Van Vuurens' stocking rate but had scattered their animals, lost hundreds to starvation.

- Consider various ways you can use available forage more efficiently—for example, increasing paddock numbers with temporary fencing, or by herding within existing fencing; gaining additional grazing through the use of portable water troughs, supplementing poor-quality forage, and so on.

- If you must destock, the sooner you know it and act, the less you have to cut.

The Economics of Destocking

The graphs in figure 2-15 illustrate why destocking earlier enables you to keep more animals. They describe what happens to 500 cows in a 160-day winter season when forage reserves fall twenty-five percent short. Normally the herd would consume 80,000 animal-days of forage, or ADs, during that time.

What plans make most sense when you estimate you'll have only 60,000 ADs standing at the onset of cold weather?

Nutrition during Slow- or Nongrowth Periods

Earlier we looked at the many advantages of greater stock density during the growing season, which you can get by increasing paddocks, amalgamating herds, and so on. The nongrowing season poses a slightly different challenge, as there are more factors to consider.

Let's say you have enough standing forage after the first freeze of autumn to last you until spring, including plants that cure well enough to supply nearly all the protein and minerals your livestock and wildlife require. Nevertheless, though all animals can find enough bulk throughout the season, its nutritional value drops steadily under continuous or rotational grazing. Both wild and domestic animals will select the best first, getting more than they need during the first half of winter and less during the second. You can buy supplements to make up the difference for the livestock but wild grazing animals will go short.

Other nasty things can happen too—such as increased poisonous plant danger in the spring, as many of the most dangerous species green up early. For the same reason, given few paddocks, cool-season grasses and forbs

suffer intense early grazing and, especially in milder climates where plants only go dormant for a very short time, overgrazing.

Grasses with better curing properties, which make them a higher-value feed for livestock, give a reddish hue to an autumn in America or a dry season in Africa or Australia. By late winter, however, these grasses have generally been grazed down and the straw yellow, less nutritious, plants are all that remain. This drop-off in feed value tends to be worse on high-rainfall leached soils, as even growing vegetation offers less in protein and minerals and more in fiber. In the tropics certainly, and probably in other environments, feed value falls off dramatically once grass is frosted.

In figure 2-16, graph A assumes supplementation will be necessary after exactly half the forage has been selected, and the need will increase. The stock, being scattered, foul a wide area, causing inefficient grazing, and are not allowed the stimulation of a move onto fresh ground.

Real situations are never neat, of course, and require close monitoring of both stock and forage, but the principle still holds. The stock have more than they need early in the season and less than required at the end. The mining industry calls this "high grading"—taking out the very best ore first. At some point the ore that remains is too low grade to pay for mining it. Miners can't make any more ore, just as you can't grow any grass in some environments and very little in others over winter. But by mixing high and low grades, they can economically mine much more. The more paddocks you have, the better you can achieve the same thing by intelligently planning the use of the forage through the entire nongrowing season and drought reserve period.

Planning for the nongrowing season grazing so that when animals enter a new paddock they also have access to previously grazed paddocks enables them to better balance their diets. (While plants are dormant, you don't face the same overgrazing dangers so you can manipulate grazing and animal impact more easily.)

Graph B in figure 2-16 shows the effect of using ten paddocks one after another. The cattle spend twenty days in each paddock over a 200-day dormant period. This scheme has several advantages. Stock do occasionally

Destocking Schedule

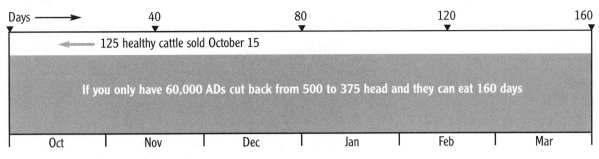

500 animals x 160 days = 80,000 AD   but 60,000 AD ÷ 160 days = 375 animals

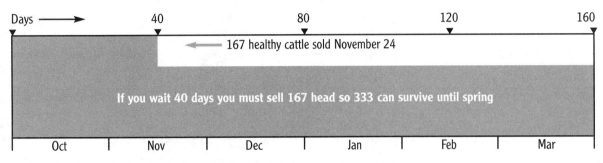

60,000 AD − (500 animals x 40 days) = 40,000 AD ÷ 120 days left = 333 animals

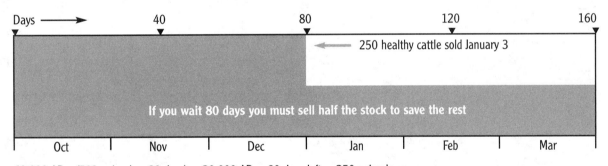

60,000 AD - (500 animals x 80 days) = 20,000 AD ÷ 80 days left = 250 animals

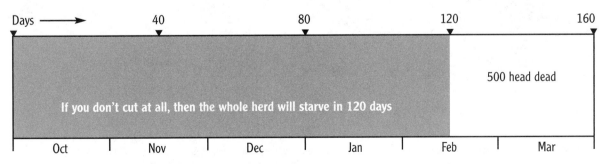

60,000 AD ÷ 500 animals = 120 days before everything is gone

Figure 2-15. If you have to destock, the sooner you do it, the less you have to cut.

move to fresh ground. The ungrazed paddocks will provide something for wildlife right to the end of the season. And even though forage value will not meet minimum daily requirements for the last ten days in each paddock, the carryover from the first ten days helps reduce the need for excessive supplementation. On the other hand, fouling will still cut grazing efficiency, and you won't eliminate supplements completely because the "rumen shock" of going from a depleted paddock to a fresh one every twenty days could reduce digestive efficiency significantly without supplementation.

People who follow a rigid rotation schedule do not usually change in winter. A three- to four-day grazing period would look like graph C in figure 2-16. This scheme would in fact keep stock moving onto fresh ground and greatly reduce the fouling problem. You might get by with minimum supplementation through the second grazing in each paddock. You would not meet the requirements of wildlife, however, and the supplement bill during the last pass through the paddocks would be high.

Suppose, however, you have 100 paddocks, and the herd stays only two days in each. Then the pattern looks like graph D in figure 2-16. Your stock move to fresh ground every other day. Forage use becomes extremely efficient. And the need for supplements is minimized. Remember, this diagram shows you the principle. In your actual plan if you had some low-lying areas prone to frosting you would be taking all the grazing that you could out of those prior to frosting. That would delay the animals coming into the frost-free paddocks, and now you would balance out the diet in those through the remaining season.

These examples are, of course, purely theoretical. Other considerations may count much more in your planning. Some paddocks may, as mentioned, be prone to

Balancing Nutrition in the Non-Growing Season

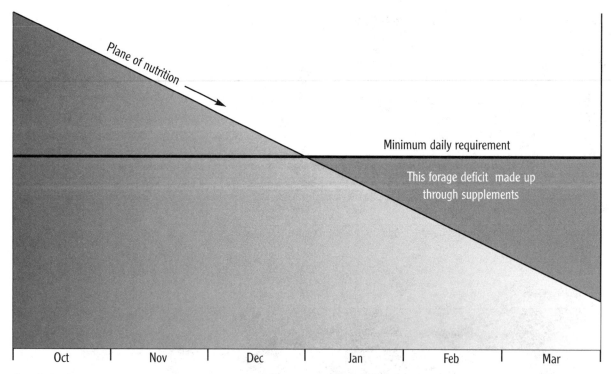

Graph A: Continuous grazing creates a large nutritional deficit at the end of the season.

Figure 2-16. The nutritional value of forage drops steadily during the nongrowing season. Animals will select the best first, getting more than they need during the first half of the nongrowing season and less during the second. You can avoid this problem to a large extent if you keep animals moving through the paddocks, minimizing the number of times they have to return to already grazed paddocks. The more paddocks you have, the fewer times animals need to graze them again, and thus the more even the plane of nutrition.

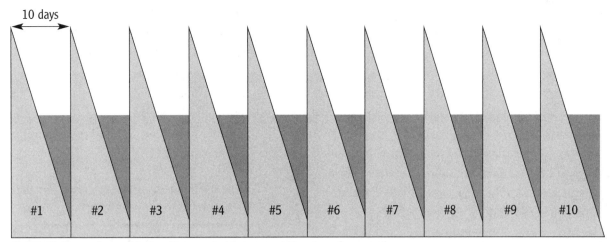

Graph B: Even a few paddocks will somewhat balance nutritional needs across time.

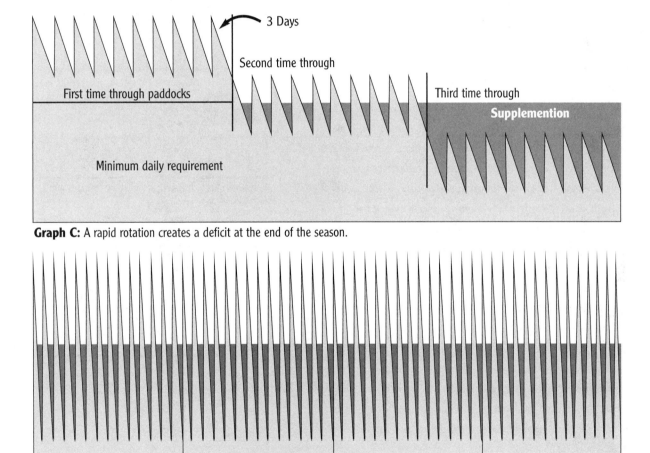

Graph C: A rapid rotation creates a deficit at the end of the season.

Graph D: High paddock numbers keep nutritional plane high throughout.

Figure 2-16. *Continued*

frosting or snow cover. Some may provide cover or a convenient place for monitoring calving or lambing. Some areas may need to be avoided at certain times because of high predation, and inevitably your plan to minimize fire risk or forage loss from fire necessitates grazing certain areas earlier in the nongrowing season. Nevertheless, amalgamating herds in the nongrowing season means more paddocks, more density, better nutrition, better mineral cycling and improved land health, and greater efficiency. (The latter applies to the growing season too, of course.)

The policy of reserving time (in ADs) versus reserving area (in ungrazed paddocks) merits another look. If you experience a drought even worse than you planned for and are forced to make a pass through all paddocks yet again, your stock will face a depleted nutritional selection without doubt. But your animals will always be better off than if you had planned for the same situation by reserving an ungrazed area.

If a drought does not occur, the accumulated grass is generally not a liability in a low-rainfall brittle environment, even when the "time reserve" planned runs a year or more. The old grass can in fact be remarkably nutritious, especially if growing on highly mineralized soils. In a high-rainfall brittle environment, however, even six-month-old grass, especially on poor, leached soils, can be of low nutritive value, so planning for drought in terms of time not area becomes doubly important.

Managing a Drought

In drought-prone areas you know that the chances of experiencing a drought are high; thus you always assume the growing season will be a poor one, either because spring rains fail to materialize or early rains are all you get. A time or drought reserve will get you through the former, but you will need to replan right through to the anticipated start of the next growing season in the case of the latter.

The point at which you replan is a matter of individual judgment based on historical rainfall records. What you should never do is wait until the end of the growing season before taking action.

At the first pang of worry about the prospects ahead of a growing season without rain, estimate your ADs of standing forage to get an idea of how long you can last in a worst-case scenario. Review all the measures you might

take that could help pull you through (see the earlier section "If You Run Out of Feed"). Then weigh your options. If you normally run several herds in separate grazing units or paddocks, you may be able to avoid buying feed or destocking if you amalgamate your herds and keep them moving through all the paddocks you have (see fig. 2-17). The upcoming section on "Single versus Multiple Herds" covers the practicalities.

How Joining Herds Affects Grazing and Recovery Periods

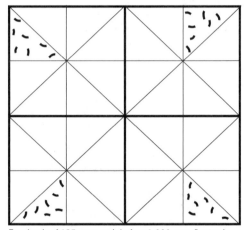

Four herds of 125 cows each in four 1,000-acre (hectare), 8-paddock grazing units.

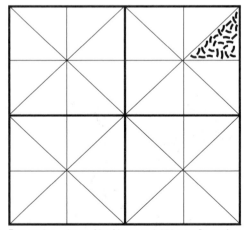

Four grazing units combined into one 4,000-acre (hectare), 32-paddock grazing unit.

Figure 2-17. In the top example, there are four herds of 125 cows each in four 1,000-acre, 8-paddock grazing units. Density is one cow per acre. If you wanted a ninety-day recovery period, grazing periods would average thirteen days. In the bottom example the grazing units have been combined into one 4,000-acre, thirty-two-paddock grazing unit. Density is now four cows per acre. Given a ninety-day recovery period, grazing periods would average three days.

Combining herds and treating several grazing units as one gives land longer recovery periods and additional benefits from greater stock density. It may also help you maintain a constant level of nutrition, thus reducing the need for supplements.

Managing Wet Winter Grazing

When too much water is the problem in the slow- or nongrowing season, as it often is in nonbrittle northern Europe where the ground doesn't freeze solid, overtrampling can occur in low-lying areas or where grass roots are shallow. (It can also occur in the growing season.) The problem can be minimized by planning grazing moves so that animals avoid areas that become boggy under wet conditions, favoring paddocks or areas on higher ground. One or more of these higher-ground paddocks may need to be designated sacrificial paddocks that change from year to year. Good land planning (see part 4) can also help address this challenge by identifying sacrificial corridors that can be used in wet conditions for movements into and out of paddocks. By planning adequate recovery times, overgrazing during the growing season will be minimized, and perennial grass roots will grow deeper, holding the soil in place and mitigating the damage from trampling in nongrowing wet periods, as highlighted in box 2-7.

Watering Large Herds

Lack of water is the most common reason people give for minimizing herd size. That's why it is so important to plan for large water supplies as you create your land plan. In most cases this will involve building larger storage reservoirs and/or piping water from two or more weak points to create one stronger one.

In the meantime, if you are faced with a drought and need to amalgamate your herds there are temporary solutions to watering one large herd, such as transporting water to the animals. Or, if large supplies are only available at some points on a ranch, you can drive animals to these points every second or third day, which has been done successfully on many ranches.

Even breeds that did not originate in arid zones can adapt to a surprisingly long time between drinks and cover

---

Box 2-7. Overtrampling and How We've Addressed It

Wet winters and the trampling damage that results are a real challenge for us here in the UK and Ireland, and as a result most people keep their animals inside in winter (a practice that is legally mandated in some countries). I've always assumed that poor grass root structure, which is common here due to overgrazing, exacerbated the problem and that if we could improve root structure trampling damage would decrease and more farmers could outwinter their animals. Two farmers, one in Scotland and another in Somerset (southern England), showed it was possible after building up a dense root structure under Holistic Planned Grazing. They are now outwintering their animals with little fear of damage from trampling.

Some years ago, when we first become a Savory Network Hub, I received a call from a farmer who said that in his first winter under Holistic Planned Grazing he moved the livestock as frequently as every two hours because the poaching [overtrampling] was so bad. The next summer he had a terrible crop of docks [weeds]. He simply carried on with the planning, and the following summer he had the most amazing grass he'd ever seen—dense, tall, and diverse. So you may need to accept some overtrampling in the beginning because it can't be helped, but if you persist with the planning you'll get something absolutely amazing.

Last year one of our farmers in Scotland ran out of winter feed when spring arrived late. Christopher, my husband, and co-hub leader, suggested he choose the paddock that could really benefit from high animal impact and feed them there. He did, and after eight hours he thought the paddock was destroyed. But three days later, after a little sunshine, he was shocked to find that paddock growing the best grass on his farm.

— *Sheila Cooke,*
Savory Network Hub Leader,
United Kingdom

SIDEBAR 2-3

The Barlite Case

On the Barlite Ranch near Marfa, Texas, 1983 proved a severe drought in which most ranches were forced to destock drastically. The manager faced reduction of the 1,200 cattle that were divided among seven grazing units, containing a total of 101 paddocks. Early in the growing season the manager had started buying in and storing hay, but as the drought progressed he grew increasingly worried and called in Allan Savory to see if there was any way to avoid destocking.

Having learned from his experience with the Van Vuuren brothers (see sidebar 2-2, "The Van Vuuren Case"), Allan suggested he combine the seven herds into one and move them through all the paddocks, which would maximize animal impact and minimize the risk of overgrazing. The two of them did some calculations and found that even if the Barlite only had an inch (25 millimeters) of rain in the remaining season they could grow at least another ounce of grass per square yard (28 grams of grass per square meter). That added up to a great tonnage of feed over the ranch as a whole.

With one herd and 101 paddocks the cattle were able to move daily. This gave each paddock a maximum dose of dung, urine, and trampling, followed by 100 days of recovery. "I didn't think anyone could move that many cattle every day," the manager said later. "But they were so used to the fences already, they pretty well moved themselves."

Finally, in October and November an inch and a half of drizzle blessed the Barlite. The manager figured it grew them 21 million pounds (9.5 metric tons) of feed. The sudden lushness of their ground stopped abruptly at their boundary fence—beyond which their neighbors' cattle had grazed continuously at half the Barlite's stocking rate. While the neighbors continued to destock, the manager bought 206 cow–calf pairs at distress prices and cut their supplemental feed bill by $26,000.

much more distance to water than generally expected of them. However, they must be worked into a routine and must never be disappointed by not finding sufficient clean water when they are brought to it. If the water point they are counting on is dry when they get there, they will tend to hang around waiting for every drip that comes and may stop grazing altogether. There are undoubtedly countless

creative ways ranchers have found to meet this challenge when so much is at stake. Given a little creativity, you can meet it too.

Creating Herd Effect

Animal impact is the tool of choice for offsetting the adverse effects of partial rest in any environment. It is achieved either through stock density, which involves a mathematical relationship between the number of animals and the size of the grazing area, or herd effect—the hoof action of excited animals on plants and soil—which is a matter of herd size and animal behavior. In nonbrittle environments stock density is often all that's needed to deal with partial rest. But areas capped with moss and lichen usually require herd effect to break it so that grass can grow. Herd effect may also prove useful in most environments for reducing brush encroachment or weed invasions or healing gullies.

In brittle environments, partly because ranches or pastoral areas tend to be large, stock density alone, unless pushed to extremes with fencing, cannot overcome partial rest. Herd effect is what's needed, and in regular doses. And the bigger the herd, the better the herd effect.

This is not a linear relationship. A herd of 1,000 can generate much more than ten times the amount of herd effect produced by 100 head. Very small herds can only achieve herd effect over small areas of land at a time. Where a herd of 1,000 is good, a herd of 2,000 is better, and a herd of 10,000 to 20,000 is even better.

The bigger the herd, the better the herd effect.

In Holistic Planned Grazing, the idea is to anticipate the areas where you will apply herd effect for any number of purposes, including the following:

- To suppress brush directly by breaking it down and opening up dense thickets
- To return oxidizing, ungrazed plant material to the soil as litter so it can decay
- To promote tighter spacing between plants, holding more litter in place and thus causing

more water to soak in rather than running off or evaporating from the soil surface

- To soften the banks of gullies and start plants growing in eroding areas or cropland being returned to pasture

- To reduce infestations of noxious weeds by direct impact and by creating soil conditions that favor fibrous-rooted grasses over tap-rooted species

- To clear firebreaks or roadsides

In the past, the presence of large numbers of pack-hunting predators in the world's brittle environments ensured large herds and more or less continuous high herd effect at some point over most grasslands.

Unfortunately, both wild and domestic animals tend to spread and remain calm when free of that danger. Driving livestock with cracking whips or dogs obviously causes herd effect but at an unacceptable price in lost animal performance and handling qualities.

The most successful methods so far for inducing herd effect involve simply herding livestock to a grazing plan without the use of fencing, which is fortunate when having to deal with vast areas, traditional pastoralists, and places where fencing is undesirable, such as national parks and other areas where wildlife is adversely affected by fencing. On fenced properties, other methods for inducing herd effect include the following:

- Using a movable electric wire to strip-graze small areas of land within larger paddocks

- Bunching the animals for even part of the day using horses and dogs

- Attracting the herd to anything that excites them (see box 2-8)

Increasingly, ranchers in the United States are adopting calmer livestock handling methods, such as those developed by animal handling specialist Bud Williams. These offer great promise in enabling us to bunch animals without stressing them, to increase animal impact through herd effect wherever desired.

Box 2-8. Using Attractants to Create Herd Effect

Attractants remain the most practical alternative in nonherding situations for creating herd effect to heal eroding gullies, thin dense brush thickets, and a variety of other site-specific, landscape-enhancing objectives. However, the downside of using attractants is the disappointingly small area actually impacted—basically, the zone around the attractant.

Consider the following examples:

- Supplements such as hay or cake fed on the ground will quickly gather and excite any herd trained to expect a handout.
- Salt will gather a herd that has been denied it for some time. Granulated livestock salt, simply fed on the ground, works best.
- Diluted molasses sprayed on weeds or firebreak areas will stimulate both grazing and herd effect on specific locations.
- Static inducements, such as salt blocks and liquid mineral licks, do not produce herd effect. Animals visit them singly and tend to loiter. Putting mineral supplements on a trailer that can be moved from place to place works better but commonly falls short of the ideal.

Occasionally situations arise that allow use of herd effect using these same attractants but only to concentrate the animals long enough to remove the bulk of the forage. Say, for example, you needed a firebreak. This can be created by spraying dilute molasses or saline solution over the vegetation, or, to open up dense brush to allow grass to begin growing, by throwing in supplement blocks cut into pieces.

Training plays a large role in the effectiveness of using attractants. Animals that have never tasted molasses, for instance, will not recognize the smell and may ignore it at first. Livestock will quickly learn to come to a whistle, though, if it consistently means a treat. Such training not only helps in stimulating herd effect but also simplifies the business of moving stock to new paddocks. Holding back a few trained animals to mix in with untrained stock vastly speeds the training.

Single versus Multiple Herds

Once you fully understand the ramifications of time, herd size, stock density, plant growth curves (see fig. 2-6), and increased energy flow to the land's productivity, you will think twice before you form an additional herd. Each additional herd carries "hidden" but real costs that can be enormous in terms of your profitability and in the rate of improvement of your land.

If you must pull bulls or heifers out of the herd at some point, consider running them on continuous graze in a sacrificial paddock that changes from year to year. This will limit the loss in productivity, because it leaves the bulk of the paddocks available for the main herd to be moved through them at greater stock density, with shorter grazing periods and longer recovery times. As a result, any precipitation you receive is more effective, and forage production is enhanced. If you try this ensure that the paddocks under continuous grazing (where plants will be overgrazed and soils overrested) are changed next season to ensure the plants in those paddocks are properly grazed and the soils once again impacted by the animals.

Single Herd Management

The greatest obstacle to reaping the benefits of combining herds into one is worry about handling so many animals and classes or species together. "Can I water that many animals at a time?" "What will happen if I put my heifers in with the bulls, steers, and cows?" "What will happen if I run horses, cattle, sheep, and goats in one herd?"

Specialized breeding programs and other considerations may make separate herds necessary, but simple numbers usually don't. The doubters generally do not believe that animals can learn behavior that makes herd size almost irrelevant to the question of handling. Or that the layout of handling facilities and the calmness of handling are far more important than herd size.

Running multispecies herds is routine for some folks, and cattle herds that include all classes of stock (bulls remain in the herd year-round, and calves are weaned without being separated from their mothers) have been run successfully for years.

Once cows or ewes are pregnant, why remove the bulls or rams from a herd? If you want calves or lambs to fall over a two-month period, then regulate it through your culling program, not by removing herd sires. Any young heifers that get bred too early can also be culled.

How do you handle single-sire breeding without many herds? One method that worked well in Namibia many years ago involved 500 registered Simmental cows running in a single herd with single-sire bulling. Not only was the rancher able to specify which cow went to which bull, he had exact conception dates on all cows, a record of each bull's performance, and no breeding of young heifers. Because each bull only served the cows once, the rancher could run up to ninety cows per bull. And fewer bulls meant he could afford to buy better bulls. Part 4 describes how he did it using a radial fencing layout with specially designed facilities at the radial's center.

That method worked well for that rancher, who was determined to keep both his single-sire bulling and a single herd, to maximize the improvement on his land. Despite the fact that he had to feed the bulls over the two-month breeding period, this strategy proved very profitable because he did not suffer the loss of forage or reduce the effectiveness of his precipitation, which would have resulted if he'd split into twenty herds. Management was eased, and the investment in fewer, better bulls more than paid for the feed required.

Use your creativity in working out challenges such as this. It is your life, your ranch or farm, and your holistic context. Don't let old beliefs, peer pressure, or fear handicap you. Given the cost in lost production that runs parallel with the number of herds, use care when checking such decisions for alignment with your holistic context. Remember that although production per animal is important, production per acre or hectare is more important if you are serious about profit.

Multiple Herd Management

That said, there may be times when running two or more herds does make the most sense and actually passes all the context checks. Box 2-9 illustrates a method you can use to help determine your herd strategy.

Wild Grazers and Browsers

The timing of livestock moves will affect wildlife in a variety of ways. You should, of course, plan to ensure that your stock avoid areas where game species are rutting or fawning, or only lightly graze areas where thick cover is

Box 2-9. Deciding the Number of Herds

You have 600 steers on 3,000 acres or hectares divided into fifty-six paddocks, averaging fifty-four acres (or hectares) in size. You'd like to push 100 steers ahead and market them early. Consider four strategies:

1. Keep all 600 together in one herd, and select the best 100 at sale time.
2. Separate the 600 into two herds (of 500 and 100) that graze through all of the paddocks but are separated in time by the recovery period being used. This allows the 100 head a greater ability to select their ideal diet, and all paddocks still benefit from the larger herd.
3. Separate the 600 into two herds in two grazing units and plan to graze the 100 steers in the grazing unit made up of the twenty best paddocks and the remaining 500 steers through the other grazing unit of thirty-six paddocks.
4. Let the 100 make the first selection in each paddock, and let the 500 remaining follow right behind with no break between the one herd and the next. (Note: Both herds must run in the same unit. Instructions for planning this option—follow-through grazing—can be found in "Grazing Two or More Herds in the Same Grazing Unit" in appendix 3 at the end of this book. Only attempt it after you have several years' experience in planning and an abundance of paddocks.)

If you desire a ninety-day average recovery period, the following factors will bear on your decision: length of grazing period, stock density, and ADA(H) harvested.

Consider the following:

- Land and thus overall production benefits from larger herds, high density, and short grazing periods.

- Livestock benefit from rapid moves to fresh ground and lower ADA(H) take.

To choose, think first about impact on the land and overall production, then on livestock.

- From the land's point of view, strategy 1 gives the lowest grazing pressure in terms of ADA(H) harvested per grazing period, the highest stock density, and the shortest grazing period. This will undoubtedly grow the most forage and make rainfall most effective. The second-best alternative is strategy 4.
- From the cattle's point of view, strategy 4 will give the best results on the 100 early-marketed steers, but not on the main herd of steers, which would be slightly better off in strategy 1.
- Strategies 2 and 3 are poor from the land's point of view because forage production and rainfall effectiveness will decrease significantly.
- From the cattle's point of view, strategy 2 is likely to be better than strategy 3 for both herds because both herds are able to better select their diets, and the large herd is moving through more paddocks more quickly.

The Calculations

Average Grazing Period

Strategies 1 and 3: Average Grazing Period = $\frac{Recovery\ Period}{\#\ Paddocks - 1}$

Strategy 2: Average Grazing Period = $\frac{Recovery\ Period}{(\#\ Paddocks \div \#\ Herds) - 1}$

Strategy 4: Average Grazing Period = $\frac{Recovery\ Period}{\#\ Paddocks - \#\ Herds}$

Average Stock Density

$= \frac{Number\ of\ Animals}{Average\ Paddock\ Size}$

Average ADA(H) per Grazing

$= \frac{Average\ Grazing\ Period \times Herd\ Size}{Average\ Paddock\ Size}$

Possible Strategy		Avg Grazing Period	Avg Stock Density	Herd Size	Avg ADA(H) Per Grazing
1.	1 herd	1.6	11.1	600	17.7
2.	2 herds	3.3	1.8	100	6.1
	All paddocks	3.3	9.3	500	30.5
3.	1 herd/20 paddocks	4.7	1.8	100	8.7
	1 herd/36 paddocks	2.6	9.3	500	24.1
4.	2 herds on follow-through	1.7/1.7	1.8/9.3	100/500	3.1+15.7 (18.8 total)

needed to protect the young of ground-nesting birds. And there are steps in the planning procedure to remind you.

But what about wildlife that also graze or browse the same paddocks your livestock do in the growing season and pay no heed to your carefully timed grazing and recovery periods? In the early days when we first planned the grazing for large livestock herds on grasslands where wild herds were also abundant, we found the wild herds quickly linked up with the domestic ones. This was possible because the fencing was simple—one or two strands of wire—and desirable from the wildlife's point of view because the forage one or two moves behind the livestock was resprouting and highly nutritious.

Initially this was a concern because it was altering planned grazing and recovery periods. A two-day grazing with a ninety-day recovery would become an eight-day grazing followed by eighty-two days of recovery. The eight-day grazing came from the two days cattle were grazing, followed by four days of inadequate recovery, then another two days of wildlife grazing/browsing—making it effectively eight days that plants were exposed to animals. In cases where livestock herds were large and animal impact high, so many new plants established that the number of plants overgrazed by the wildlife was not a problem. It could potentially be a serious problem if the impact supplied by the livestock herd was minimal. We don't yet have enough experience with large wildlife herds running in conjunction with stock to give more advice than this: make every effort to maximize animal impact while minimizing stock time in a paddock; anticipate that wild grazers/browsers will associate with your herd in the growing season, grazing behind it on the regrowth; monitor carefully and keep us posted.

Matching Animal Cycles to Land Cycles

Ranchers and farmers may argue forever about the best time of year for calving and lambing. Many variables govern the decision—markets, parasite hatches, seed and awn problems, weather, and more. That said, however, tradition often outweighs most other factors. If profit is important to you and you aim to lessen the burden of supplemental feed, low conception rates, and care of young stock, checking your herd's varying nutritional demands against the natural cycles of forage might prove interesting.

Traditional practice in the United States usually dictates that calves come in early spring to be weaned and sold in the fall. If the market looks bad, some may be held over until the next spring. But if other factors (say the absence of screwworms) permit it, why not consider summer calves? Generally, the economics of holding them over and selling them at eighteen months look much better than for calves dropped in early spring.

Once again, the standard worksheet provides a handy form that allows you to compare several strategies or classes of stock and different kinds of country. Once you start considering possibilities, the decisions can become extraordinarily complex and might be best sorted out with the gross profit analysis check. In that case, you would treat each possible breeding and production policy as a separate enterprise and compare it with the alternatives, taking into account the costs associated with each policy and the anticipated income at varying risk levels.

Roughly speaking, cows need a rising plane of nutrition for conception and above-average feed for six months of lactation. They need less when dry and pregnant. Sheep and goats follow a similar pattern except that they're usually not lactating and pregnant at the same time. The graphs in figure 2-18 show the general nutritional needs of cattle and sheep.

Rethinking the cycle of forage may even turn up the fact that keeping stock through the winter doesn't pay at all. Or you might want a combination of operations so that market times support a constant flow of cash.

The point is that when you actually begin to do your grazing planning for plants, animals, and soils, no tradition need be sacred. And as your land improves, new possibilities will open—for instance, more cool-season grasses and/or an improved water cycle extend the grazing season, fibrous plants give way to those higher in energy and protein, poisonous plant problems fade, and water supplies increase.

A handbook of this nature has to stick to principle rather than detail because the latter varies so much from

Nutrition and Forage Cycles

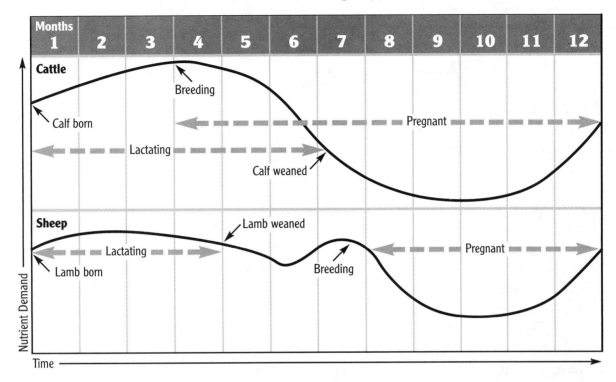

Forage Availability

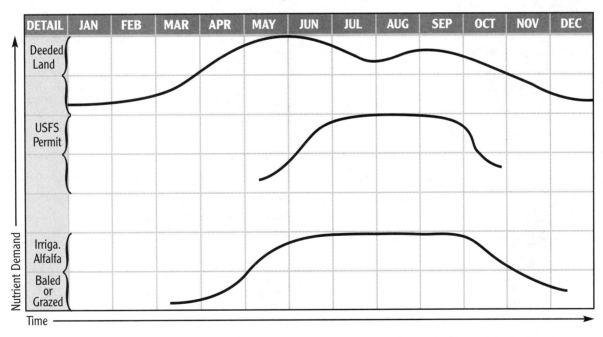

Figure 2-18. The top graph shows that nutritional requirements for both cattle and sheep are high when breeding and that both need to be on a rising plane of nutrition prior to breeding. The aim is to match those requirements to the forage ability shown in the graph below. Should you run summer calves and early lambs, or ewes and yearlings on the U.S. Forest Service permit? If south slopes and low-elevation parks were not grazed late in summer, would elk winter there instead of on alfalfa? Could you sell a better hunt? Could you substitute grazing for the last alfalfa cut? Finish lambs on alfalfa?

ranch to ranch, farm to farm, region to region, and year to year. In working the details out in practice, you will want to consult others. But beware. There is a great deal of expertise available on animal nutrition, animal production strategies, and more, and you will be pressured by your friends, neighbors, and even some of the experts themselves to heed it. Remember, though, that you are the only expert on the whole you are managing—which includes your family, your ranch/farm, and your animals. So be careful to check all decisions for alignment with your holistic context, especially those informed by a narrow area of expertise.

Pests, Parasites, and Other Challenges

The sequence in which paddocks are grazed can affect a host of considerations besides the productivity of plants. Most of these relationships are too case specific to detail here, but the principle is simple: you can probably manipulate any situation characterized by a strict routine by changing your own routine, if you find the critical point.

Here are some examples:

- Many parasites, such as liver flukes, leave their livestock host for part of their cycle. Records will show where your herd was grazing at this critical period. Planning to have your animals elsewhere when the parasites again need a host may nearly eliminate the problem.

- Some pests either breed in manure or seem to spend a lot of time on moderately fresh manure. Moving animals frequently onto fresh ground will keep them ahead of emerging young flies and may leave some adults behind.

- Livestock losses rise when predators have good cover, insufficient wild prey, and young to feed.

- Plan to calve, lamb, or kid, on safe sites when natural prey is plentiful and predator needs are low.

- Predation is always higher on livestock spread thinly over large areas of land and reduced with larger herds and mixed herds accompanied by imprinted dogs or llamas, for example.

- When floods occur in predictable seasons, plan to graze affected areas afterward, since ungrazed plants will slow the flow and catch sediment.

- Many poison plants threaten stock only for limited periods or when other forage is scarce. Graze these areas at safe times or take out very light grazings in susceptible times.

- Ground-nesting birds and other animals may require certain habitat only for nesting or breeding. Keeping stock out of these areas at critical times and while heavy cover is required by the young birds should increase their numbers.

- Grass fires may threaten certain areas during the dormant season, but grazing them early may cut the danger if they do burn; every day your animals were on them is a day saved on later grazing elsewhere.

The number of options that must be sorted in the process of making a good grazing plan can obviously overwhelm even the most able intelligence that tries to cope without a method for organizing and displaying information. Fortunately, such a method exists.

Creating Your Plan

H*OLISTIC* P*LANNED* G*RAZING* is done according to an aide memoire (French for memory aid) because, as the textbook explains, the task involves more variables and complexity than the average mind can handle. As the term applies here, an aide memoire differs from a simple checklist because it gives a sequence for making small decisions that takes into account the effect of one decision building on another. All the factors you have to consider—dry water points, fires that have burned half the paddocks, poisonous plants, physiological state of the animals, ground birds nesting, and so on—are dealt with in a logical order. You don't have to worry that something else should have come first.

The aide's main benefit is that it forces you to concentrate on one step—one small piece of the larger puzzle—at a time, but in the full knowledge that the pieces will build to the best possible plan. As you complete each step, wipe it from your mind and focus on the next. Nothing gets left out so relax and enjoy yourself.

The years of experience embodied in the aide memoire assure that you will incorporate all the concerns you may have, even if they don't turn up in the order you anticipate. Don't skip ahead. Along the way you will decide where to apply the tools of rest, grazing and animal impact, or technology to create the landscape your holistic context requires. The management guidelines for time, stock density, herd effect, and population management (covered in detail in *Holistic Management*) will assist you in preventing overgrazing and applying animal

impact where needed. Other steps will enable you to plot the application of other tools and uses of the land. As you constantly monitor your progress, you'll be able to modify and replan according to changing circumstances.

Holistic Planned Grazing includes two aide memoires—one for the growing season and one for the slow- or nongrowing season—because planning needs and aims differ depending on the season. In either case, the aide memoire will generate a plan that takes into account any number of variables and displays them in a graphic way and in great detail. Thus you can advance several priorities at once, see your situation at a glance, and change the plan easily as circumstances demand.

Guidelines for Planning

As in the case of the financial plan, the instinctive human fear of planning in general is your greatest enemy. More than you wish to admit, however, this fear reflects the old "what you don't know won't hurt you" syndrome. Planning will show up areas where you can't attain the ideal—and that hurts. Specifically remember the following:

- Follow the aide memoire carefully, moving through the steps in the order they appear. Concentrate only on what you are asked to do in each step.

- *Plan on paper*, specifically the Grazing Plan and Control Chart, because with planned grazing you will be focusing on the recovery periods (rather than grazing periods), which can only be seen on a chart like this. There are any number of other factors that can prove overwhelming unless noted somewhere, and others cannot contribute much to a plan they can't see.

- Keep your first plans *simple*—minimize the number of herds and avoid sophisticated grazing strategies, such as those covered in appendixes 3 and 4, until you have mastered the basics.

- Plan *pessimistically* on any point on which you have any doubt at all. This will help you avoid unpleasant surprises later on.

- Plan *creatively* every time. Easy planning year after year will tempt you to abandon the process and fall into a routine that will sooner or later cost you plenty.

- Create one plan per herd for the *growing season* and one plan per herd for the *nongrowing season* and drought reserve.

SIDEBAR 2-4

When Herding Livestock without Fencing

Over the past decade, the Africa Centre for Holistic Management (a Savory Network Hub in Zimbabwe) has learned much about herding livestock to regenerate large tracts of land and wildlife habitat using predator-friendly practices in the midst of abundant predators (lions, leopards, African wild dogs, etc.). They have developed a modified aide memoire for use in such situations, which, because they have an unlimited number of virtual paddocks, eliminates the need for calculating a range of grazing and recovery periods and keeps the arithmetic simple. The aide is still paired with the same planning chart presented in this handbook.

The Africa Centre has also worked with the Savory Institute to develop and test a simplified planning procedure and a simplified chart for agropastoralists who graze over common lands, and a training program and materials that address the unique challenges in this context. These materials are available through the Savory Institute. Training programs are currently available through the Africa Centre and will soon become available through other Savory Network Hubs.

- Your *aim* in the growing season is to *maximize forage* production.

- Your *aim* in the slow- or nongrowing season is to *ration out the forage* grown in the previous growing season, and return any old, standing forage to the ground where it can be incorporated and enhance mineral cycling.

- *Monitor the plan.* Remember that no plan ever goes exactly to plan. What you expect to happen rarely does, and thus planning is always a process of planning, monitoring, and controlling, or adjusting. *Reread the step on implementing your plan* (step 14, growing season aide memoire; step 13 nongrowing season aide memoire) *often.* And refer back to this step whenever you are unsure of what you're observing, when you experience a drought, or when you run into trouble of any kind.

The Aide Memoires for Holistic Planned Grazing

Holistic Management requires you to plan, monitor progress continuously, control deviation as soon as possible, and replan whenever necessary. Even though this plan–monitor–control–replan sequence proceeds without gaps and covers emergency situations, livestock operations usually call for major planning twice a year.

The planning generally takes a person two to three hours to complete. Implementing the plan is of course a continuous process, and it will always require adjusting. The first 12 or 13 steps in the aide memoires (the growing season aide memoire has one more) are the actual planning steps. They are preceded by some important preplanning activity and followed by guidelines for implementation and monitoring and recording the results of each season's plan for use in future planning.

Make your first plan at least a month before the onset of the main growing season. Over the growing season planning period you will be trying to grow as much forage as possible, and you do not have to plan to a specific date. The plan remains *open* because, while you are steadily building up the volume of forage, you don't know when growth will slow or end.

Make the second plan toward the end of the growing season, when forage reserves available for the nongrowth period become known. During this planning period, you ration out the forage over the months ahead to a theoretical end point, which should be a month or more after your most pessimistic estimate of when new growth could occur. This additional "month or more" becomes your *drought reserve*. And because you're planning to a specific date this nongrowing season plan becomes a *closed* plan.

Figure 2-19 illustrates the open-ended and closed plan principle, showing how the time reserve for drought fits in. In some regions where rainfall is very low and unreliable, there will be a major overlap of the drought reserve in the closed plan and the potential start of the next growing season due to the necessity in some environments of making drought reserves extend to a year or more.

The growing season and nongrowing season aide memoires follow. The figures referred to within them appear at the end of this chapter.

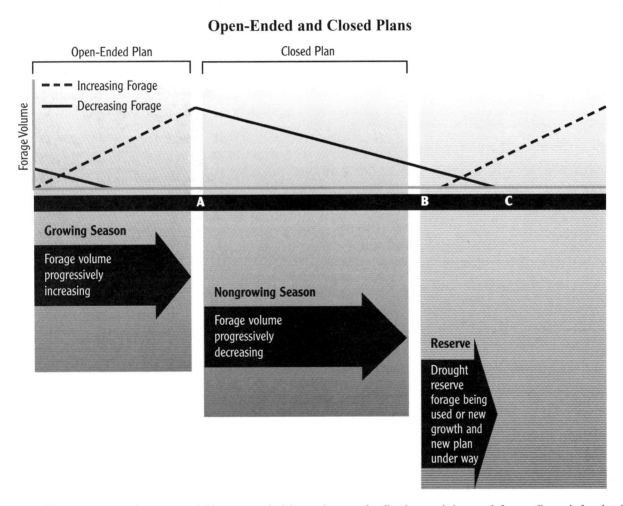

Open-Ended and Closed Plans

Figure 2-19. The growing season plan is open ended because you don't know when growth will end or exactly how much forage will grow before that date. The nongrowing season plan goes into effect once growth stops (point A). It is a closed plan, in that you are rationing out a known amount of forage over a specific period of time (from A to B), which should extend a month or more later than your most pessimistic estimate of when new growth could occur. This additional time (from B to C) is the drought reserve.

Growing Season Aide Memoire

Reminders
- Use this aide memoire in conjunction with the grazing plan and control chart.
- Follow the steps in order, as each step builds on the previous one.

SIDEBAR 2-5

The Steps in Brief

Step 1. Set up the grazing chart.

Step 2. Note management concerns affecting the whole grazing unit.

Step 3. Record herd information.

Step 4. Record livestock exclusion periods.

Step 5. Check for unfavorable grazing patterns.

Step 6. Record paddocks still available.

Step 7. Note paddocks requiring special attention.

Step 8. Rate paddock productivity.

Step 9. Determine the length of recovery period(s).

Step 10. Calculate average grazing periods.

Step 11. Calculate actual grazing periods.

Step 12. Plot the grazings.

Step 13. Make a final check of your plan.

Step 14. Implementing and monitoring the plan.

Step 15. Keeping the record.

- Concentrate on each step. Record your response on the chart, then wipe that step from your mind and move on to the next.
- The aim of the growing season plan is to grow the maximum amount of high-quality forage.
- The best possible plan for now will develop from your knowledge.
- Plan afresh every time. Easy planning year after year may tempt you to abandon the process in favor of a routine that will lead you into trouble.

Preplanning

Meet as a planning team to discuss the growing season to be planned. Answer the following questions:

- Is the entire area going to serve as one *grazing unit* with one herd, or will different areas become different grazing units for separate herds? Your answer will determine how many plans you need to create. Remember: the fewer the herds the better.
- What stocking rate do you intend to carry? This could change in the course of the planning, but at this stage you'll probably have an idea of what you want and what you think is reasonable.
- What crops will be planted, and where? (Use a worksheet, as shown in fig. 2-20, to schedule planting, cultivating, and harvesting and to plan crop rotations, and so on, and keep it on file. You will use it for both the growing- and the nongrowing-season plans).
- What other factors will influence your planning—wildlife, poisonous plants, other uses on the land, times when critical people are away, and so on? Make a list and keep it handy.
- Has your ecological monitoring (see part 3) identified any problems that need to be resolved? Do you need to increase animal impact in some areas? Increase rest in other areas? Reduce partial rest all over?

Note: The map illustrated in figure 2-21 was used in completing all the charts illustrated in this chapter. Refer to the two completed charts at the end of this chapter for an example of how to record steps 1 through 7 (fig. 2-22) and steps 8 through 13 (fig. 2-23).

Step 1. Set Up the Grazing Chart

Use one chart per grazing unit. If there are more paddocks than lines on the chart, cut and join charts to provide enough.

- Record the year, the name of the grazing unit, and that this is a growing season, or open-ended, plan.
- Write in the names of the months for which you are planning in the blank columns at the top of the chart. Each month has fine lines that demarcate thirty-one days. If a specific month has fewer days than that, draw a pencil line down the column through the extra day(s).

- Record the paddock numbers in both column 3's (on the left and right) and the paddock sizes in the left column 3.

- If crop fields will have livestock on them record them as paddocks.

Note: The growing season plan is an open-ended plan. This means that you can terminate the plan on any day you choose well into the growing season because forage will generally be increasing. You can then commence planning from that point as you see how the season progresses.

Most of the time, you only run one herd per grazing unit and use one chart per grazing unit or herd. If you are running a smaller herd in a sacrificial paddock within the grazing unit, make sure you change the paddock each year. (See "Single versus Multiple Herds" in the first section of part 2 for details.)

Step 2. Note Management Concerns Affecting the Whole Grazing Unit

Using color-coded felt pens, box in all the management events that could affect your animals or management, no matter what paddock the animals will be in at any time.

- Draw vertical lines through all the paddocks on the starting and ending dates of livestock management events: breeding (red), calving/lambing/kidding (blue), weaning (yellow), and connect them across the top of the chart. Note the title of the event on the top connecting line.

- Do the same for other factors identified in your preplanning: hunting seasons, wildlife factors, family vacations, and so on.

Step 3. Record Herd Information

- Record the types of animals and their numbers each month in rows 29 to 33. Convert total stock numbers to SAUs.

- Record the total SAUs for each month in row 34. If they change over the months, note the peak figure in the first column, as shown in figure 2-24.

- Enter the total grazing unit size in row 35 (over on the left) and the stocking rate (grazing unit size divided by peak SAU) in row 36 (lower left corner).

Note: If you need more rows than available on the chart, you can paste on rows from another chart. But it may be simpler to just use the many columns to the right, making sure your total SAU figure is the total of all animals.

When converting stock numbers to SAUs use the method you are most comfortable with as long as you are consistent each year. Several methods are suggested in the first section in box 2-1, "Conversion to Standard Animal Units."

Step 4. Record Livestock Exclusion Periods

- Using a fine, color-coded pen, draw a horizontal line through any time periods when a paddock *cannot under any circumstances have the herd in it*—for example, because of lack of water, annual flooding, hay cutting, crops growing, trout spawning, birds nesting, poisonous plants, wildlife factors, timber plantings.

- If you plan to remove herd sires to run in one continuously grazed, sacrificial paddock, draw a line though that paddock over those months because the main herd will be excluded.

- Use a different color for each factor, and explain the meaning of the color with a legend in the "Remarks" space at the bottom of the chart. Do not write on the main body of the chart.

Step 5. Check for Unfavorable Grazing Patterns

If this is your first year to plan, move on to the next step. Otherwise, look over past grazing charts and clearly mark (with coded colors) any paddocks in which the following has occurred:

- Paddocks repeatedly received an *H* early or late in past growing seasons (exclude the paddocks affected entirely for the early and late periods this season).

- Paddocks received inadequate recovery time in the last growing season due to the end of growing conditions, such as the first frost (avoid these paddocks early in this growing season).

- The growing season has started early, leaving you with drought reserve forage in some paddocks (plan to remove the old growth before it affects new growth or livestock condition).

Note: An adverse or dangerous grazing pattern can develop where any paddocks are heavily grazed at the same time of the biological (not calendar) year during early and/or late growth two years consecutively. Rows 21, 22, and 23 on past grazing charts will help you determine early and late growth times.

- Look for paddocks that received an "H" (for heavily grazed) early or late in past growing seasons. You are trying to avoid grazing a paddock heavily more than twice in a row during early growth when plants are growing on root reserves. And you are looking for paddocks grazed heavily late in the season more than once without adequate time for nutrient translocation before going dormant.

- Look for paddocks with an *L* (for lightly grazed) late in the nongrowing season, and plan to have them grazed early in the growing season before too many oxidizing plants affect new growth and animal performance. These paddocks are often those with leftover drought reserve forage. The paddocks you are marking now are not unavailable, as in step 4. You just want to avoid them at certain times.

Step 6. Record Paddocks Still Available

In row 26 under each month note the number of paddocks still available. If a paddock is available for more than fifty percent of the month, count it as available for the whole month.

Note: As you gain experience, count a paddock as available if the unavailable period is less than the average recovery period you are likely to use. For example, if a paddock is marked unavailable for fifty days and your recovery periods are likely to range from thirty to ninety days (an average of sixty days) count that paddock as available. In this way you can keep livestock moves as short as possible without sacrificing recovery periods. When you later plot the animal moves, this unavailable period will become a recovery period and you will not put the animals in that paddock over that time.

Step 7. Note Paddocks Requiring Special Attention

Focus on each paddock in turn, and, using a color-coded highlighter that you can later write over with pencil, clearly mark through any paddock that needs special treatment over any particular time period (see note below).

Explain the meaning of the highlighter colors with a legend in the "Remarks" space at the bottom of the chart. Do not write on the main body of the chart.

Note: Factors to consider might include the following:

- Bare, capped, or eroding ground that needs healing or noxious weeds you want to reduce—both of which could require herd effect

- Areas you want to rest more to create brush or other cover conditions for wildlife

- Areas in which you want to remove the bulk of forage toward the end of the growing season, as part of a fire prevention plan

- Paddocks you plan to strip graze

At the same time, think about any limitations that could affect your stock or require management attention—proximity to crops or a neighbor's bulls, poisonous plants, sharp seed heads that could harm lambs, boggy areas that need to be avoided, lack of shelter or adequate water, parasite cycles that need to be broken, and so on.

Review the list of planning factors you created earlier. Note any wildlife management factors and crop preparation plans. And do not forget to consider multiple-use factors that could affect a paddock, such as logging operations or ecotourism activities and the like.

Step 8. Rate Paddock Productivity

- Rate productivity based on each paddock's forage quality and size relative to all the other paddocks. Use the method you are most comfortable using (see below).

- Record the rating for each paddock in column 2, and the average of all ratings at the bottom.

Note: To avoid large variations in forage nutritional value between paddock moves it is important to rate the productivity of each paddock relative to all others so that adjustments can later be made to minimize those variations. If any paddock is withdrawn for the entire period, such as a sacrificial paddock, do not include it in these ratings.

There are two ways to rate paddock productivity, which is based on forage quality and paddock size. Box 2-10 describes a simple rating using a 1–10 scale that will help you get started. Once you have some experience, use ADA (or ADH) ratings.

Using ADA or ADH Ratings

If you have more than a year of past planning use the ADA(H) figures from column 8 in your last closed plan.

- Transfer these "actual" figures into column 1.

Because this is the approximate amount of forage per acre or hectare, you now need to factor in the size of each paddock.

- Multiply the ADA(H) figure in column 1 by the size of each paddock (column 3) to give you the approximate animal days of forage available in each paddock – this becomes the paddock rating.
- Enter the paddock rating in column 2. (If this number is very large divide it by 1,000, so that it easily fits in the column, as shown in the figure 2-25.)

Step 9. Determine the Length of Recovery Period(s)

Give considerable thought to this step to ensure minimal overgrazing of plants.

- Use one recovery period when you have many paddocks per herd, or are strip grazing or herding within a few paddocks. Use two recovery periods (minimum and maximum) where you do not have many paddocks and are neither strip grazing nor herding in them.
- Record your decision in row 27 under each month as either one recovery period or two (e.g., ninety, or thirty to ninety days).

Box 2-10. Rating with the 1–10 Scale

If you are just beginning, use a simple 1-10 scale to get an approximate quality rating for each paddock. Pose the question, "Which paddock would hold my stock for the longest time?" Rate that paddock a 10. Then rate all the other paddocks relative to that one. A paddock in which animals could stay half as long would rate a 5 and so on.

- Record the rating for each paddock directly into column 2.

Note: Remember that plants can be overgrazed during a grazing period that is too long, or after a recovery period that is too short for root system recovery to have taken place. This is why, when using fencing and few paddocks, it is wise to plan on two recovery periods over the growing season—one shorter recovery period to use over times of rapid daily plant growth, and a longer recovery period to use when growth is slow. When you later plot the grazing periods you will use the longer grazing and recovery periods and through monitoring will drop to the shorter grazing periods when appropriate. The faster the growth, the shorter the recovery period; the slower the growth, the longer the recovery period required. Here are some guidelines to help until you gain your own experience:

- For a range of recovery periods
  - On arid and semiarid rangeland, thirty to ninety days will usually suffice to minimize overgrazing. Given higher or more effective precipitation and more fibrous vegetation, twenty to forty or sixty days may do.
  - For pastures (irrigated or not), particularly those with runner-type grasses, try fifteen to thirty days.
  - In areas that receive prolonged periods of adverse growth conditions you may have to further lengthen recovery periods to 150 days or more.
- For a single recovery period
  - If you have many paddocks, you can use one recovery period as long as the subsequent grazing periods average no more than about three days. For example, if you had 100 paddocks and used 180-day recovery periods to achieve only one grazing on average through a growing season, the grazing periods would average 1.8 days.
  - If you have few paddocks but are either strip grazing or herding within paddocks, use one recovery period of the length you determine wise. The grazing periods will be longer than desired, and if the animals are allowed to wander it will result in plants being overgrazed. This is avoided by making sure the animals are not allowed to graze any area in the paddock for more than about three days. This can be accomplished either by using strip grazing with portable wire or by herding.

The preceding guidelines are to get you started. Once you have gained experience you may choose to modify them. Depending on your goals for your land and livestock there will be times when you may use far longer recovery periods. For instance, you might want to develop

---

SIDEBAR 2-6

If You Have Very Few Paddocks . . .

Although the ideal is to plan a range of recovery periods, and thus grazing periods (in step 11), when you only have three or four paddocks the grazing periods become so long that plants are going to be overgrazed whatever you do. But the *overgrazing will occur in the grazing period, which is less detrimental than the overgrazing that occurs following an inadequate recovery period.* So feel free to plan with just a single recovery period, using the longest—that is, ninety days if your range is thirty to ninety days—and in step 11 one grazing period. Try to get paddocks up to ten or more as quickly as your financial planning and the weak link in the livestock enterprise allow.

---

maximum grass root depth and volume to hold more water and carbon in the soil, knowing that livestock performance would drop due to less frequent moves to fresh grazing.

Step 10. Calculate Average Grazing Periods

If you have one recovery period in row 27, calculate one average grazing period.

- To get the average grazing period, divide the recovery period (row 27) by one less than the number of paddocks per herd (row 26).
- Record average grazing periods in row 28.

If you have two recovery periods in row 27, calculate both minimum and maximum grazing periods.

- Average minimum grazing period (AMGP) = minimum recovery period (row 27) divided by number of paddocks (row 26) – 1.
- Average maximum grazing period (AMxGP) = maximum recovery period (row 27) divided by number of paddocks (row 26) – 1.
- Record AMGP and AMxGP in row 28 (e.g., 3.4 – 6.8), and do not round off decimals.

Note: If your paddock numbers change over the season, due to cropping, hay cutting, and other operations, a commonsense solution is to calculate the grazing periods based on the average number of paddocks available over the time when paddock numbers vary.

If you wish to plan follow-through grazing, where two or more herds run in the same grazing unit, with one herd entering a paddock as another leaves that paddock, see appendix 3. This is tricky to plan, and you should have considerable experience before you attempt it.

Step 11. *Calculate Actual Grazing Periods*
Calculate the actual grazing period(s) for each grazing.

- Divide the rating of each paddock in (column 2) by the average of all ratings at the bottom of column 2.
- Then multiply this figure by the average grazing period in row 28, and this time round off decimal points to the nearest day.
- Record the actual grazing period planned for each paddock in column 4.

Where there are two average grazing periods in row 28 (because paddock numbers change), calculate the actual grazing periods for both and record the answers in column 4 with a slash between minimum and maximum grazing periods. If you have very great differences in the lengths of these final grazing periods, and you have few paddocks, check to ensure you minimize overgrazing of plants.

Note: Where grazing periods show extremes from very short to very long, add up all the paddock minimum grazing periods, and then perform the three following steps:

1. From the total subtract the longest minimum grazing period to find the actual recovery periods for these paddocks.
2. If any recovery period is much too short, you must add days to the minimum grazing periods in other paddocks that can absorb them, as shown in the example in figure 2-26.
3. Follow the same procedure for maximum grazing periods, though the problems will probably be less critical if you cannot make complete adjustments.

Step 12. *Plot the Grazings*
Look over all the color-coded events on the chart and plan where you will actually place the animals based on where they need to be at certain times. Have a map of the ranch or farm handy so you can confirm that the planned moves are practical. Plan the moves backward or forward as needed to get animals to the right place at the right time.

Use a soft pencil to mark in the actual moves of the herd on the chart. Show the length of the grazing period (taken from column 4) by the length of the penciled line.

Note: This is an important step. If you are not very experienced, follow these guidelines:

- Events such as calving, lambing, or kidding show clearly on the chart if they occur in the planned months. In planning the moves over such times, you will want to prevent separation of mothers and young. You can do that by moving animals to adjacent paddocks so that gates can be left open and mothers can return to hidden young until they are ready to bring them into the herd.
- Plan grazings backward from periods when livestock nutrition or other needs are critical. To do this, reserve paddocks that you know will have high-quality forage and/or ample cover at critical times, then plan which paddocks animals should come from in order to get there. Planning forward, or worse, merely rotating animals, rarely assures this. Animal performance consequently suffers, or livestock may conflict with wildlife or other land uses.
- Where you have problems or special management concerns in some paddocks, as indicated by your color-coded marks, you need to address them. For example, in paddocks with a poison plant danger, you might cut grazing times to the minimum period so that grazing pressure will not force consumption of the problem plants. Here the advantage of planning on a chart becomes apparent because every day taken off a *single* grazing period actually takes a day off the recovery period in *every* paddock.

This cumulative effect on the recovery periods, which can have disastrous consequences, will always show up on a chart, but will not show up in a notebook or on scraps of paper.

- Most failures experienced to date have been because managers either stopped planning and reverted to a simple rotation or did not watch recovery periods on the chart. They monitored grazing periods only, rather than the actual recovery periods, to determine whether changes were needed. Both of these practices are dangerous.

- If you see that ground-nesting birds or other wildlife need cover soon after a grazing, you might also want to drop to the minimum grazing period in that paddock.

- If you are using two grazing and recovery periods and you drop to the minimum grazing period in several paddocks, keep an eye on the recovery periods. You may have to lengthen grazing periods in other paddocks that can take them to ensure recovery time in all paddocks is adequate.

- Use the longest (maximum) grazing period given in column 4 so that your plan is conservative. The greatest danger of overgrazing stems from recovery periods that are too short, not grazing periods that are too long. When in doubt, slow down.

- As you plot the grazings, keep an eye on the recovery periods—the actual recovery periods will show up on the chart.

- Remember that in creating an open-ended plan you simply plan the grazings until well into the growing season, knowing there will tend to be more forage available as you proceed through these months. If the season turns out well you simply keep extending the plan.

If the season turns out to be a dry one and growth stops, close your open-ended growing season plan and immediately create a closed, nongrowing-season plan.

Step 13. Make a Final Check of Your Plan

If you have concerns about forage availability, do a field check (see note below) in the paddocks you are most concerned about.

- Using SAU figures from row 34 look at each plotted grazing and calculate the amount of forage required in ADA(H), and record the figure in pencil to the left of the plotted grazing.

- If any of the ADA(H) figures look worrying, simply go out to those paddocks and do random field checks. If they show forage may be short, then make adjustments to the plan by reducing days in those paddocks and adding them to other paddocks. The number of days reduced in each paddock must be added to some other paddock or recovery periods will become too short and plants will likely be overgrazed.

Note: In making a field check, you should check several sample areas in any paddock of concern to see if they could feed one animal for one day. To get the sample area size if you are working in acres, divide 4,840 (square yards to an acre) by the ADA figure to the left of your penciled grazing line and punch the square root button on your calculator. To get the sample area size if you are working in hectares, divide 10,000 (square meters to a hectare) by the ADH figure to the left of your penciled grazing line and punch the square root button on your calculator. This square root figure will give you the length of each side of the square, in yards or meters, needed to feed one animal for one day in that paddock.

As you look at sample areas in any paddock pose the question, Would this area support one animal for one day? It is important that you factor in the physiological state of the animals at this point by qualifying your question. If they are lactating cows, ask, Would this area feed one cow very comfortably today? If lactating cows are about to be bred and you wanted them to be on a rising plane of nutrition you could ask, Would this area feed one cow very comfortably and with forage to spare? Or if you are dealing with a lot of wildlife grazing also modify your question to ask, Would this area feed one animal for one day but leave plenty of forage for the wildlife?

Congratulations! Your plan is now complete.

Step 14. *Implementing and Monitoring the Plan*

In implementing your growing season plan follow these guidelines:

- If you are using two grazing/recovery periods you must monitor the daily growth rate of the plants (see box 2-11).
  - Move livestock as slowly as nutritional needs and the maximum grazing period permit when growth rates are slow.
  - When growth becomes rapid, drop to the minimum grazing period. Once growth slows (prolonged rapid growth is uncommon) return to the maximum grazing period.
  - Note growth rates for each five-day period every month in row 23. Use *S* for slow, *F* for fast, and *0* for no growth.

- If the season is going well, extend your plan at any time, but if you are well into the season and it is turning out to be a poor one with little rain, don't just hang on hoping. Start your nongrowing season plan early as a precaution.

- If you suspect you have misjudged the quality of a paddock, move the animals immediately and note on the chart that the paddock was less productive than thought. If you suspect animals may run out of forage in other paddocks (and that you might be overstocked), do a stocking rate check out in the field (see box 2-11).

- Monitor to ensure that cover, feed, lack of disturbance at critical times, and any other wildlife needs noted on the chart, are being met.

Step 15. *Keeping the Record*

Record what happens so you can fine-tune your paddock assessments and management decisions for future planning.

1. Record actual events *in ink* as the season progresses. For every paddock, ink in a line on the chart that covers the number of days the herd actually spent there. (See figure 2-27 for an example).

2. Behind or to the right of the actual grazing line, record the volume of forage taken by the animals in ADA(H), and then follow this figure with your assessment of how heavy that grazing was. Use *L* for light (you can hardly see the paddock was grazed), *H* for heavy (you had to remove animals), and *M* for medium (everything in between). Do not judge close to water, but well out in the paddocks. Your record would look something like this: 34/M if you wanted to show that the grazing over the days marked, was 34 ADA(H) and was a medium grazing in terms of total bulk.

3. Clearly flag, or mark, any paddocks where serious errors were made—for example, quality was not as good as planned, or grazing periods were too long or recovery periods too short—so the same mistakes will not be made in the next plan. Also note any paddocks that had noneffective recovery periods due to grazing pattern and season. Remember that you have based paddock assessments on ADA(H) estimates or actual performance in previous seasons (with adjustments). In brittle environments especially, errors are certain to occur in your estimates. You will find that livestock at higher densities in smaller, better-defined areas, are going to show up assessment faults better than you could have determined them by any other means. Therefore, keep an open mind and make note of all errors that can affect the next plan.

4. Record precipitation received in rows 21 and 22. The spaces between the heavier lines represent five-day periods. You may want to note the exact day of major storms with a dot on a single line and a written comment in the "Remarks" section on the chart. Note your average

Box 2-11. Monitoring Notes

Monitoring Daily Growth Rates

When you have many paddocks, as you will always have when herding without fencing, or when strip grazing within paddocks using temporary electric fence, you do not need to monitor daily growth rates because all grazing periods will be short relative to the recovery period—often as long as an entire growing season. (You do need to continue monitoring livestock condition, forage bulk, wildlife factors, and other concerns that affect your plan.)

With fewer paddocks, you always plan minimum and maximum grazing and recovery periods and monitor daily growth rates so you can drop to the minimum when growth is fast, and return to the maximum when growth slows.

To monitor daily growth rates, compare grazed plants in the paddock the animals have just left to grazed plants in the paddock the animals are now in to help you judge the rate of regrowth. You can also mark grazed plants in any paddock with flagged wire stakes or other markers to help you observe regrowth. Alternatively, you can place portable wire cages over some plants and later compare them to the grazed plants surrounding them. A plant has recovered when it looks like its ungrazed neighbor. Bear the following points in mind:

- Don't assume the daily growth rate is fast just because good rains have fallen. Factors other than rain, such as temperature, existing soil moisture, and effective rainfall, also affect growth rate. Immediately following rain plants may continue to grow slowly due to waterlogged soil and poor aeration, and only several days later, when aeration improves, start growing rapidly.
- If you have dropped to minimum grazing periods because growth is rapid remember that growth rates will seldom be rapid for more than a few days at a time. As the growth rate slows, move back toward the longer (maximum) grazing period shown in column 4. If you ever have doubts about the growth rate, assume it is slow—as it commonly will be. If you are very observant, you will find the color of the grass also indicates whether or not growth is rapid.

- If the growth rate continues to be rapid for a prolonged time and the grazings therefore shift far off the plotted moves, you will need to replan from that date forward. This replanning normally only involves changing the plotted moves.

Minimizing Overgrazing

- With low paddock numbers, overgrazing can occur either when stock stay too long in a paddock during fast growth or when they return too quickly during slow growth.
- *With higher paddock numbers the greatest danger is in returning too soon during slow growth.* This rarely happens when paddock numbers are as high as 100 or more, which is often the case when herding, but you still need to keep an eye on the recovery period in each paddock if for any reason you have to shorten grazing periods.
- If you are using a range of recovery periods and growth rates are slow, move livestock as slowly as nutritional needs and the maximum grazing period guideline allow. Animals generally perform better when moving through a paddock faster, but if they move too quickly, plants will be overgrazed. If you can see ahead that animals are going to return to a paddock before plants have recovered, you are moving too fast, and you must slow down. Not doing so (generally because you want to favor your animals) is one of the most common mistakes people make, with costly consequences.

If You Run Out of Forage

- If you are using a range of grazing periods and find that animals run out of forage in any paddock, even though you are using the minimum (shortest) grazing period, you have seriously misjudged the paddock. If you run out of forage in many paddocks, your judgment could again be at fault, but you are more likely to be overstocked.
- If you have many paddocks using a single grazing period and recovery period and run out of forage in any of the paddocks, the reasons will be the same.

Box 2-11. *Continued*

- If you suspect paddock quality was misjudged when you initially planned, move the animals immediately and note on the chart that the paddock was not as good as you thought it was. If you suspect that animals may run out of forage in other paddocks (and that you might be overstocked), do a stocking rate check out in the field.

Field Check for Stocking Rate

Select the paddocks you are most concerned about, or some at random. To determine the size of your sample areas in each paddock, multiply the minimum grazing period listed in column 4 by the SAU figure in row 34,

to get the animal days of forage that would be taken out during the shortest grazing. Then divide the total by the size of the paddock (column 3) to give you the ADA(H) for that paddock. Divide 4,840 (or 10,000 for metric) by this ADA(H) figure to get the number of square yards or meters needed to feed one animal for one day. Push the square root button on the calculator to get the length of each side of the square to sample. *If the sampled areas in most of the paddocks will not feed one animal for one day, you are definitely carrying too many animals and will need to reduce their numbers quickly.*

annual precipitation in row 37. Record the total precipitation received for the season in row 38.

5. Add any comments on the chart that will help you do better in future planning—not on a piece of paper that can get lost, but on the back of the chart where you have plenty of space.

6. When you judge that the growing season has ended, and you are ready to move to a closed plan, draw a brown line down through all paddocks, and label it "Growth Ended."

Nongrowing Season Aide Memoire

Reminders

- Use this aide memoire in conjunction with the grazing plan and control chart.

- Follow the steps in order, as each step builds on the previous one.

- Concentrate on each step. Record your response on the chart, then wipe that step from your mind and move on to the next.

- The aim of the nongrowing season plan is to ration out the forage through the nongrowing and drought reserve months, while maintaining the best plane of nutrition you can for your stock and your wildlife and preparing the land for the following season.

- Although some plant growth may take place in the nongrowing season, it is generally slow and you are unlikely to overgraze plants.

- The best possible plan for now will develop from your knowledge.

- Plan afresh every time. Easy planning year after year may tempt you to abandon the process in favor of a routine that will lead you into trouble.

SIDEBAR 2-7

The Steps in Brief

Step 1. Set up the grazing chart.

Step 2. Note management concerns affecting the whole grazing unit.

Step 3. Record herd information.

Step 4. Record livestock exclusion periods.

Step 5. Record paddocks still available.

Step 6. Note paddocks requiring special attention.

Step 7. Rate paddock productivity.

Step 8. Plan the drought reserve.

Step 9. Plan the number of selections and the average grazing periods.

Step 10. Calculate actual grazing periods.

Step 11. Plot the grazings.

Step 12. Make a final check of your plan.

Step 13. Implementing and monitoring the plan.

Step 14. Keeping the record.

Preplanning

Meet as a planning team to discuss the nongrowing season to be planned. Answer the following questions:

- Is the entire area going to serve as one *grazing unit* with one herd, or will different areas become different grazing units for separate herds? Your answer will determine how many plans you need to create. Remember: the fewer the herds the better.

- What stocking rate do you intend to carry? This could change in the course of the planning, but at this stage you'll probably have an idea of what you want and what you think is reasonable.

- What crops will be planted, and where? (Use a worksheet, as shown in figure 2-20, to schedule planting, cultivating, and harvesting and to plan crop rotations, and so on, and keep it on file. You will use it for both the growing- and the nongrowing-season plans).

- What other factors will influence your planning—wildlife, poisonous plants, hay feeding, other uses on the land, times when critical people are away, and so on? Make a list and keep it handy.

- Has your ecological monitoring (see part 3) identified any problems that need to be resolved? Do you need to increase animal impact in some areas? Increase rest in other areas? Reduce partial rest all over?

Note: The map illustrated in figure 2-21 was used to complete all of the planning charts. Refer to the two completed charts (figs. 2-28 and 2-29) at the end this chapter for an example of how to record steps 1–5 and steps 6–12.

Step 1. Set Up the Grazing Chart

Use one chart per grazing unit. If there are more paddocks than lines on the chart cut and join charts to provide enough.

- Record the year, the name of the grazing unit, and that this is a nongrowing season, or closed, plan.

- Write in the names of the months for which you are planning in the blank columns at the top of the chart. Each month has fine lines that demarcate thirty-one days. If a specific month has fewer days than that, draw a pencil line down the column through the extra day(s).

- Record the paddock numbers in both column 3s (one on the left and one on the right) and the paddock sizes in the left column 3.

- If crop fields will have livestock on them record them as paddocks.

- Change the heading of column 4 to "Nongrowth" and label the blank column to the right of it "Drought Reserve."

- If you planned the previous, just ending, growing season and have not already done so, make a brown line down that chart marking the end of the growing season and the start of this plan.

Note: The nongrowing season plan is a closed plan. This is because as you go through the season there will be less and less forage and you are moving toward an unpredictable beginning of the next growing season. This, in addition to planning constantly for droughts, necessitates taking the plan to a theoretical point of no more forage sometime after new growth can be expected.

Although running more than one herd will not carry the same hidden costs as in the growing season, fewer, larger herds will generally favor the land, and eventually water retention through better litter laying and generally higher animal impact.

Step 2. Note Management Concerns Affecting the Whole Grazing Unit

Using color-coded felt pens, box in all the management events that could affect your animals or management, no matter what paddock the animals will be in at any time.

- Draw vertical lines through all the paddocks on the starting and ending dates of livestock management events: breeding (red), calving/ lambing/kidding (blue), weaning (yellow), and connect them across the top of the chart. Note the title of the event on the top connecting line.

- Do the same for other factors identified in your preplanning: hunting seasons, wildlife factors, family vacations, and so on.

Step 3. Record Herd Information

- Record the types of animals and their numbers each month in rows 29 to 33. Convert total stock numbers to SAUs.

- Record the total SAUs for each month in row 34. If they change over the months, note the peak figure in the first column, as shown in figure 2-24.

- Enter the total grazing unit size in row 35 (over on the left) and in light pencil note the stocking rate (grazing unit size divided by peak SAU) in row 36 (lower left corner).

Note: If you need more rows than available on the chart, you can paste on rows from another chart. But it may be simpler to just use the many columns to the right, making sure your total SAU figure is the total of all animals.

When converting stock numbers to SAUs use the method you are most comfortable with as long as you are consistent each year. Several methods are suggested in the first section, in box 2-1, "Conversion to Standard Animal Units."

Step 4. Record Livestock Exclusion Periods

- Using a fine, color-coded pen, draw a horizontal line through any time periods when a paddock *cannot under any circumstances have the herd in it*—for example, because of lack of water, thatch or hay cutting, wildlife factors, paddock is burned out, and so on.

- If you have a sacrificial paddock under continuous grazing draw a line through that paddock over those months because the main herd will be excluded.

- Use a different color for each factor, and explain the meaning of the color with a legend in the "Remarks" space at the bottom of the chart. Do not write on the main body of the chart.

Step 5. Record Paddocks Still Available

In row 26 under each month note the number of paddocks still available.

Step 6. Note Paddocks Requiring Special Attention

Focus on each paddock in turn, and, using a color-coded highlighter you can later write over with pencil, clearly mark through any paddock that needs special treatment over any particular time period (see note below).

Explain the meaning of the highlighter colors with a legend in the "Remarks" space at the bottom of the chart. Do not write on the main body of the chart.

Note: This is a critical step. Take your time. Factors to consider might include the following:

- Any paddocks showing old oxidizing grass that needs trampling

- Bare, capped, or eroding ground that needs healing, or noxious weeds you want to reduce—both of which could require herd effect

- Areas you want to rest more to create brush cover for wildlife, or woodland for timber production

At the same time, think about any limitations that could affect your stock or require management attention—proximity to a neighbor's bulls, poisonous plants, sharp seed heads that could harm lambs, boggy areas that need to be avoided, lack of shelter, parasite cycles that need to be broken, and so on.

Review the list of planning factors you created earlier. Note any wildlife management factors and crop preparation plans. And don't forget to consider multiple-use factors that could affect a paddock—such as logging operations or ecotourism activities and the like.

Step 7. Rate Paddock Productivity

- Rate productivity based on forage quality and size of all paddocks relative to one another using the method you are most comfortable using.

- Record the rating for each paddock in column 6 (column 2 if using 1–10 ratings) and the average of all ratings at the bottom.

Note: To avoid large variations in forage nutritional value between paddock moves it is important to rate the productivity of each paddock relative to all others so that adjustments can later be made to minimize those variations. If any paddock is withdrawn for the entire period, such as a sacrificial paddock, do not include it in these ratings.

There are several ways to rate paddock productivity, which is based on forage quality and paddock size. Box 2-10 describes a simple rating using a 1–10 scale that will help you get started. Once you have some experience, use ADA/ADH ratings.

Using ADA or ADH Ratings

If you have planned in prior years and have the charts from those years you can now base paddock ratings on past performance, with adjustments, which is a more accurate measure of quality. Do the following for each paddock:

- Add up the ADA(H) figures from your previous nongrowing season plan—they appear in ink to the right of each grazing line—and record the total in column 8 of that plan.

- Transfer these figures to column 5 of this year's plan. Make adjustments up or down for each paddock based on how much better or worse you feel the production was in the growing season just experienced. Paddocks given an *H* (heavy) after the last grazing would probably rate lower, and so might a paddock grazed shortly before the growing season ended.

- Next, multiply the ADA(H) figure in column 5 for each paddock by the size of the paddock (column 3) and record the resulting figure in column 6. This gives you the estimated AD available in each paddock, and it can be a large number.

- Finally note the average ADs for all paddocks at the bottom of column 6, as shown in figure 2-30.

Step 8. Plan the Drought Reserve

Determine whether you will have roughly enough forage to carry your animals through the nongrowing season and drought reserve period. Use the rows on the lower right of the chart to record the following:

- **Estimated total ADs:** Add up the estimated ADs available in each paddock (column 6), and record the total in row A. (If this is your first plan, leave row A blank.)

- **Expected days of nongrowth in an average season:** Think hard about this figure and use all the past records you have to guide you. Record your estimate in row B.

- **Days of bulk feeding:** If you expect prolonged periods under heavy snow, or any other situation that would lead to bulk feeding of the animals when they are taking little if anything from the land, record the days in row C. Work out the bulk feed requirements for those days and record the type and amount of feed under the appropriate months in row 24.

- **Days of drought reserve required:** Enter the number of days you need to set aside as drought reserve in row D. This figure reflects the number of days of grazing you plan to reserve in case the next growing season is a late or dry one. In some very arid and erratic rainfall areas, the drought reserve could include most of the year.

- **Total days grazing required:** Add the figures in rows B and D and subtract the figure in row C (if entered). The resulting figure is the total days of grazing required off the land. Record this figure in row E.

- **Estimated carrying capacity:** Divide the animal days in row A by the figure in row E. This is your estimated carrying capacity—the number of animals the land can support without any additional bulk feeding. Record this figure in row F.

  If the figure in row F is much lower than the number of animals you have, you may want to reduce animals now. If you have to reduce stock, remember, the earlier you do it the less you have to reduce.

Note: Now is the time to plan for drier-than-average years. Either you have just come through a drier growing season and are thus planning for a "known" drought, or you have come through a normal season and are planning for a delayed start of the next growing season, or a "possible" drought. Either way, planning for drought is needed, and at times both will coincide, making this drought planning critical. Right now, you are doing a preliminary check on stocking rate to cover the drought you are in, or could be in soon, to see if radical adjustments to animal numbers may be required. And you are also planning the length of time that you estimate is required for the drought reserve.

Step 9. Plan the Number of Selections and the Average Grazing Periods

Each time the animals return to a paddock, they are selecting their diets from forage that has decreased in volume and quality because of the earlier grazing with no regrowth between. Thus the fewer times a paddock is grazed the better. But the length of the grazing period also counts as animals tend to become more nutritionally stressed the longer they remain in a paddock. So try to keep grazing periods short, but not so short that recovery time is insufficient for fouling to wear off.

Plan the number of selections over the nongrowth period only. Do not factor in the drought reserve time, which will lengthen grazing periods unnecessarily. If the following season's growth begins on time, animals will have performed better due to the shorter grazing periods. If the new growing season starts late you will bring the animals through all the paddocks, once again using the forage deliberately left as drought reserve.

- **Number of selections:** Divide the estimated days of nongrowth in row B by the number of selections you would like to use—try for two or three—to see what your recovery periods would be for the nongrowing season, minus the drought reserve. For example: If your total days of grazing required in the nongrowth period is 151 and you want to use three selections, that would give you 151 ÷ 3 = 50 days of recovery from fouling between the grazings. Consider using a single selection if you have 100 or more paddocks available or are herding or strip grazing within fewer paddocks.

- **Average grazing period—nongrowth:** Divide the recovery period by the number of paddocks (in a closed plan there is no need to account for the one they are in) to get the average grazing period you will use during the nongrowing season. For example, if you have fourteen paddocks, then you would divide the average fifty days recovery by fourteen to give you 3.6-day grazing periods on average. Record the average grazing period in row 28 up to the point where growth normally begins (see box 2-12).

- **Average grazing period—drought reserve:** Use the same procedure to calculate the average grazing period as above. If the figure in row D ("Days of Drought Reserve Required") was 60, and you were taking one selection and had fourteen paddocks available, the average grazing period during the drought reserve would be 60 ÷ 14 = 4.3 days. Record the average grazing period in row 28 over the drought reserve months.

Step 10. Calculate Actual Grazing Periods

Convert the average grazing periods for the nongrowing months into actual figures that take into account forage

---

Box 2-12. How Many Selections?

When plants aren't growing it is difficult to overgraze them so you are less concerned with recovery periods in the nongrowing season. You want to be sure that you keep stock concentrated for their impact, to lay litter, break soil capping, and distribute manure well, and that paddocks are given enough time between grazings for the fouling (dung and urine) to wear off. In a closed plan your focus is on rationing out the forage through the nongrowing season as well as a drought reserve period, while maintaining the best plane of nutrition for your stock that you can.

Using Two Selections

Assume you have decided on a recovery of 180 days. Divide the recovery period by the number of paddocks to get the average grazing period you will use, as in this example:

> If you have fourteen paddocks, grazing periods would average 180 days of recovery ÷ 14 paddocks = 12.8 days.

If you decided you wanted to have your animals take two selections to shorten grazing periods, but knowing no growth was taking place and there was no danger of overgrazing plants, you would first divide the total recovery period by two, as in this example:

180 days of recovery ÷ 2 selections = 90 days of recovery. Grazing periods would average 90 days of recovery ÷ 14 paddocks = 6.2 days, with adequate recovery for fouling to wear off.

Paddocks would be grazed twice for about six days, reducing the need for supplementing in the first selection. The same amount of grazing is being taken out, but more evenly, delaying the nutritional stress on the animals until later. You might have to begin supplementing during the second selection, but overall you would probably require less supplement.

Using One Selection

On the other hand, in the early years, especially when you are likely to have few paddocks, one selection might be best. In this case, other concerns, such as the need to train animals to electric fences and to gain management experience, might outweigh livestock nutritional needs. Also bear in mind that with few paddocks stock density is automatically also low, and thus fouling is not the big factor it becomes as paddocks increase and stock density rises.

Take all of the above into account in deciding which number of selections is best for your situation and your animals right now.

quality according to the paddock rating in column 6 (or in column 2 if this is your first nongrowing season plan).

- Divide each paddock rating by the average rating for all paddocks, and multiply that figure by the average grazing period for all paddocks (row 28).
- Record these figures, rounded off to whole numbers, in column 4.

Next, do the same over the drought reserve months and, for each paddock, record the drought reserve's actual grazing periods in the column you labeled "Drought Reserve" to the right of column 4.

Step 11. Plot the Grazings

Look over all the color-coded events on the chart and plot the herd moves through to the end of the average nongrowing season. Have a map of the ranch or farm handy so you will know whether the moves you choose are practical. Plan moves backward or forward as needed to get animals to the right place at the right time.

Use a soft pencil to mark in the actual moves of the herd on the chart. Show the length of the grazing period (taken from column 4) by the length of the penciled l ine.

When the herd moves have been plotted to the end of an average nongrowing season, continue plotting the moves in light pencil through the drought reserve period.

Note: This is an important step. If you are not very experienced, follow these guidelines:

- Events such as calving, lambing, or kidding show clearly on the chart if they occur in the planned months. In planning the moves over such times you will want to prevent separation of mothers and young. You can do that by moving animals to adjacent paddocks so that gates can be left open and mothers can return to hidden young until they are ready to bring them into the herd.
- Plan grazings backward from periods when livestock nutrition or other needs are critical. To do this, reserve paddocks that you know will have

high-quality forage and/or ample cover at critical times, then plan which paddocks animals should come from in order to get there. Planning forward, or worse, merely rotating animals, rarely assures this. Animal performance consequently suffers or livestock may conflict with wildlife or other land uses.

- Where you have problems or special management concerns in some paddocks, as indicated by your color-coded marks, you need to address them. In paddocks with a poison plant danger, for instance, you might decide now to reduce the grazing period so that grazing pressure won't force consumption of the problem plants. If you make this adjustment in a couple of paddocks, it will shorten the recovery periods for all paddocks, but in the nongrowing season this is not an issue.

Step 12. Make a Final Check of Your Plan

Calculate the ADA(H) you plan to take during each grazing period for the nongrowing season and drought reserve.

- Work this ADA(H) figure out using the SAU figures from row 34 for animal numbers. Multiply the SAU figure by the number of days in each grazing period, and then divide by the size of that paddock. Do this for every grazing period in each paddock through the end of the drought reserve.

Record each ADA(H) figure lightly in pencil just to the left of the plotted grazing line.

- Add together all the ADA(H) figures you penciled in to the left of the grazing line in each paddock to get the total ADA(H) you are demanding from each paddock over the nongrowth and drought reserve periods. Record the total figure for each paddock in column 7. Then check to ensure the paddocks will yield what you plan to take from them.
- Check several sample areas in any paddock of concern to see if they could feed one animal for one day.

If your samples are falling short of what is required, despite earlier checking on the stocking rate, you probably have too many animals and need to take action early.

Note: In making a field check, to get the sample area size, divide 4,840 (square yards to an acre) by the ADA figure for the planned demand from the paddock in column 7. If you are working in hectares, divide 10,000 (square meters to a hectare) by the ADH figure for the planned demand from the paddock in column 7.

This will give you the size of area (in square yards or square meters) needed to feed one animal (SAU) through the nongrowing season and drought reserve in that paddock. Push the square root button on the calculator, and that will give you the length of the sides of the square sample areas you should check in the field. After pacing out each square, ask yourself, Will this area feed one animal adequately and still leave enough to provide litter for the soil and feed for wildlife?

If the field checks show a need to reduce animal numbers, remember that doing so quickly is extremely important. *The earlier in the season you reduce animal numbers the fewer you have to reduce to carry through.* And the earlier you reduce, the more options you will have, because in such a year those who haven't planned their grazing will start selling as they begin to run out of forage. Prices will drop as the market is flooded.

Congratulations! Your plan is now complete.

Step 13. *Implementing and Monitoring the Plan*
In implementing your plan follow these guidelines:

Monitor Forage Consumption
- If your animals run out of forage in any paddock, you probably misjudged the volume or quality of the forage. Note on the chart that the paddock was not as good as you thought it was. If you run out of forage in many paddocks, you are more likely to be overstocked and must take action before animal performance is seriously depressed.

- Monitor to ensure that cover, feed, lack of disturbance at critical times, and any other wildlife needs noted on the chart, are being met.

- If the next growing season begins on time, mark a green line right down the chart through all paddocks on approximately the day you consider new growth seriously began. You can generally make a good assessment of this point a couple of weeks into growth—making sure it did actually begin and was not a false start. Remember, you are recording actual rainfall in row 21. It is not the first rain you are establishing now but the beginning of serious growth, which denotes the end of the nongrowing season—and thus the beginning of the growing season.

- If the growing season begins on time you can erase the drought reserve grazings you plotted. They will not be needed because you will be starting the next growing season plan.

Using the Drought Reserve
If the growing season does not start when expected, begin using your drought reserve, but first check to see if the moves you plotted earlier need to be plotted afresh based on the state of the land and your animals now. Remember that your drought reserve is feed that is spread over all your paddocks and left over for this period stretching from the time of anticipated growth to the last possible dates for growth.

As some growth begins, with rising temperatures or early showers, you must not speed up the moves to chase the green growth. Actively growing plants require adequate recovery periods before being regrazed, even if it means putting some stress on the animals.

Sometimes growth does not begin before the end of the drought reserve. This is unlikely to happen if your planned drought reserve was based on long-term weather records with adjustments if the weather you experience is genuinely changing. However, if you are faced with this situation then, without delay, create a new closed plan that carries you some months past the present point. As you do this you are likely to face one of two scenarios: (1) despite the drought you are now in and having planned for the latest possible new growth, you do still have forage scattered throughout all paddocks, or (2) you have almost nothing left in any paddocks.

Remember This Simple Rule

Every day animals stay in a paddock longer than they needed to, their performance will be poorer. But every day animals move out of a paddock sooner than they needed to, you lose a day of recovery time in every paddock.

Failure to understand this cumulative effect on recovery periods leads to managers favoring animals to the point that they later run out of forage entirely, which they then blame (falsely) on the "drought." So do keep to the plotted grazing periods until you know that any new flush was not just a flash in the pan. Otherwise you risk seriously overgrazing plants during the critical early growth stage.

In either case, the natural tendency of people in such dire straits, and who cannot offload animals, is to allow their animals to scatter so that individuals can fend for themselves. Although it is counterintuitive, we have learned through hard experience that more animals are saved from starvation if the animals are kept together and moved through all paddocks on very short grazing periods.

Step 14. Keeping the Record

Record what happens so you can fine-tune your paddock assessments and management decisions for future planning.

1. Record actual events in ink as the season progresses. For every paddock, ink in a line on the chart that covers the number of days the herd actually spent there. (See figure 2-27 for an example.)

2. Behind or to the right of the actual grazing line, record the volume of forage taken by the animals in ADA(H), and then follow this figure with your assessment of how heavy that grazing was. Use *L* for light (you can hardly see the paddock was grazed), *H* for heavy (you had to remove animals), and *M* for medium (everything in between). Do not judge close to water, but well out in the paddocks. Your

record would look something like this: 34/M if you wanted to show that the grazing over the days marked was 34 ADA(H) and was a medium grazing in terms of total bulk.

3. Clearly flag, or mark, any paddocks where serious errors were made—for example, an inaccurate quality rating, a clash with wildlife, or some other factor that should have been avoided—and make a note for future planning. Remember that you have based paddock assessments on ADA(H) estimates or actual performance in previous seasons (with adjustments). In erratic rainfall environments especially, errors are certain to occur in your estimates. You will find that livestock at higher densities in smaller, better-defined areas are going to reveal assessment faults better than you could have determined them by any other means. Therefore, keep an open mind and make note of all errors that can affect the next plan.

4. Add any comments on the chart that will help you do better in future planning—not on a piece of paper that can get lost, but on the back of the chart where you have plenty of space.

5. Finally, summarize livestock and land performance in the grazing unit over the year in the lower right corner of the chart. The two blank lines provided are for recording any other significant measures. Most important is to note the total yield per acre or hectare of products sold—meat, milk, wool, and so on. (See figure 2-31 for a filled-in example.)

6. If any growth at all takes place over the period planned, mark in the letter G (for growth) in row 23 in the five-day periods provided. Some greening-up or very slow growth does occur in many areas in the nongrowing season. If this growth spreads over more of each year, as it can with an increasingly effective water cycle, this record should prove useful. Significant growth is likely in some paddocks but not

others so put a G in row 23 indicating growth took place with an * alongside paddock(s) in column 3 in which growth took place.

7. When the next growing season begins—make your best guess of the actual starting date—draw a green line down through all paddocks for that day, and label it "Growth Started." At this point you would create your next growing-season plan and abandon the rest of this plan, which usually includes only the drought reserve.

8. Record the forage each paddock yielded in this nongrowing season. To get these figures, add together all the grazings noted in ADA(H) in each paddock from the green line you have just drawn, back to the brown line you drew to mark the end of the previous growing season and the beginning of the present nongrowing season. Using a light pencil, record these figures in column 8.

9. Now go through and make some adjustments to the figures you just recorded in column 8. Paddocks that you can see were grazed late in the last growing season, and thus did not fully regrow, could potentially yield more than your penciled-in figure shows. And those paddocks that have an *M* or an *L* behind the last grazing taken also had more forage in them than the ADA(H) taken shows. In either case, estimate how much additional forage each of these paddocks could yield and add this amount to the yields in column 8. Record the final figures, in ink, in column 8. These figures represent the best possible comparison of paddock quality and "performance."

Summary

THE AIDE MEMOIRE FOR HOLISTIC PLANNED GRAZING and the concepts behind it represent over fifty years of trial and error on four continents in many situations. It works. Each step builds on the next one, so the order is important. Don't skip any. If they truly don't apply to your situation, pass on—but think about them anyway. You'll find yourself in control of an amazing array of varied and subtle factors affecting your operation.

There must be something philosophically significant in the fact that of all the procedures in this book the one for grazing planning is the most orderly. Its object—the health and growth of plants and animals—is of course the most unpredictable, complex, and baffling subject that human intelligence ever sought to fathom. Without thinking this paradox all the way through, you still have to live with its significance and realize that those who don't recognize it fail.

Because living organisms never stop changing and even microbes have plans of their own, you yourself can never stop planning. If you do, you will fall into habits and routines that worked when you first planned them but will make you miss the opportunities of the ever-changing game. Eventually they will lead to ruin.

The human weakness for routine runs deep, but we've come a long way from our wild roots, and we often confuse the wonderfully dynamic patterns of nature and the reassuring regularity of machines. The greening cottonwoods in spring, the wedges of geese in fall, the grasshoppers of summer, as timeless in their regularity as they may seem, are not the same as the New York to Los Angeles Boeing overhead.

You plan for living things, not because you can ever hope to bind them to your pattern, but so you can fit yourself into theirs. Thus you must do it twice every year and sometimes more often to stay with the game.

Nothing is perfect, though, and the Aide Memoire for Holistic Planned Grazing that has evolved continuously over the years will continue to develop. Do your planning, keep track of your ideas, and if you develop improvements, let us know.

Master Cropping Plan

Date 2018 | **WORKSHEET** | Planning Sheet Column Reference Crops | Worksheet No. 4

	Jan	Feb	March	April	May	June	July	August	Sept	Oct	Nov	Dec	
Biological Year	Snow	Snow	←		Rains / ← Main Growth		→	→			Snow Possible		
Field/Size													
D2 250 ac	Livestock	→	← Field Preparation			Corn		→	Harvest	←	Livestock on Residue	→	
H3 400 ac	Livestock	→	← Preparation & Plant					→	Harvest	←	Livestock	→	
Total													

Figure 2-20. An example of a master worksheet to show fields being used for crops and when they would be available for grazing crop residues. Worksheets can be used to show many years of cropping practices.

River Bend Ranch

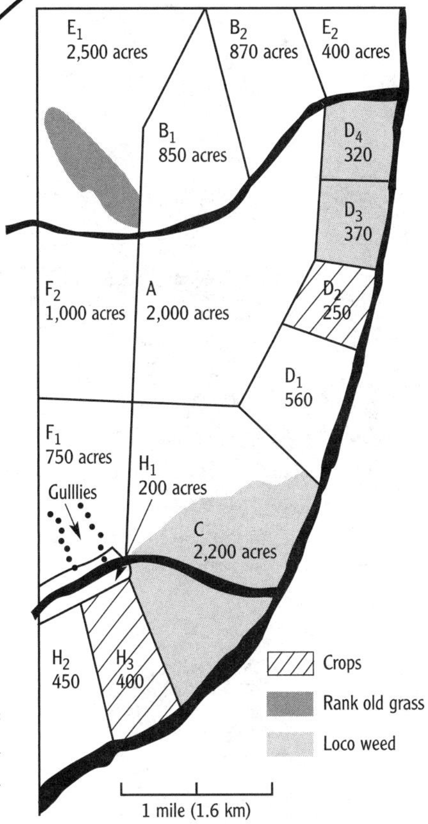

E1 2,500 acres
B2 870 acres
E2 400 acres
B1 850 acres
D4 320
D3 370
F2 1,000 acres
A 2,000 acres
D2 250
D1 560
F1 750 acres
H1 200 acres
Gulllies
C 2,200 acres
H2 450
H3 400
Crops
Rank old grass
Loco weed
1 mile (1.6 km)

Figure 2-21. Sometimes a whole ranch is planned as a grazing unit. This example is used in completing the planning charts illustrated on the pages that follow.

Sample Grazing Chart (Steps 1 - 7)

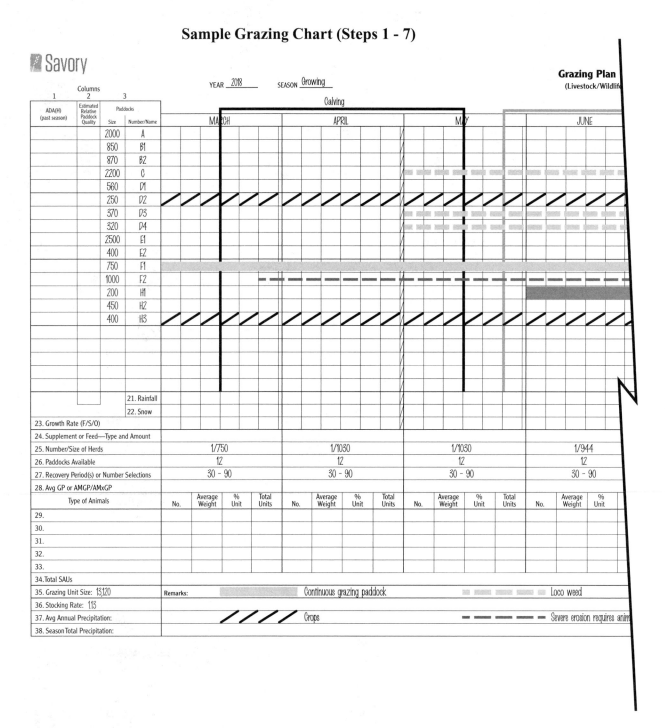

Figure 2-22. This partially completed chart shows the major management concerns in the growing season for this ranch—calving and breeding—and special concerns in individual paddocks, such as loco weed, starting in May. The yellow line running from March through September shows that this paddock will be under continuous graze for the entire season, as indicated in the "Remarks" section below. On line 25, note that in each month there will be one herd, but that herd size changes in some months. Line 27 shows that a range of recovery periods will be used—thirty to ninety days—throughout the season.

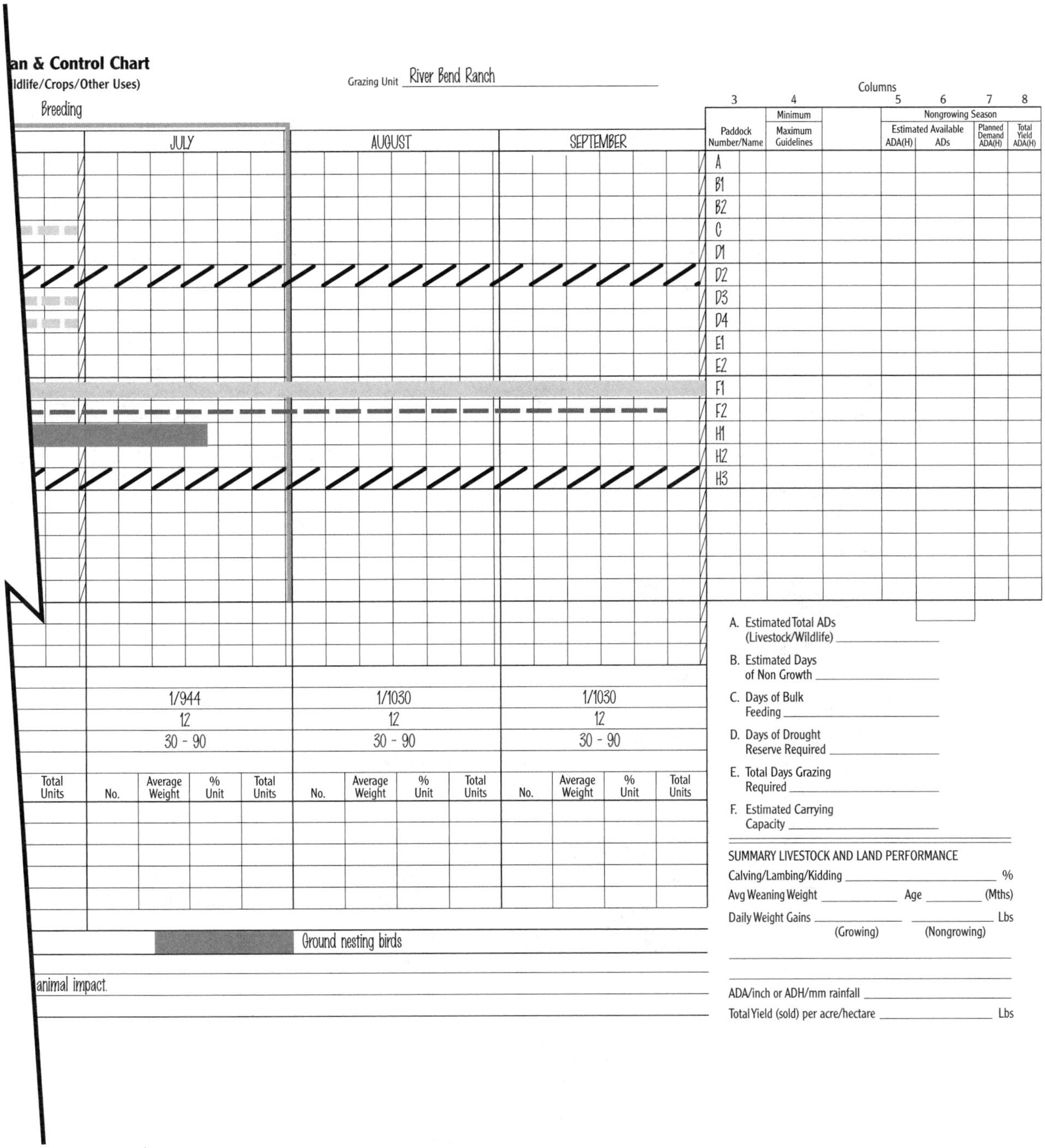

an & Control Chart
(ldlife/Crops/Other Uses)

Grazing Unit River Bend Ranch

Breeding

	JULY	AUGUST	SEPTEMBER

Columns					
3	4		5	6	7	8

Paddock Number/Name	Minimum Maximum Guidelines		Nongrowing Season			
			Estimated Available ADA(H)	ADs	Planned Demand ADA(H)	Total Yield ADA(H)
A						
B1						
B2						
C						
D1						
D2						
D3						
D4						
E1						
E2						
F1						
F2						
H1						
H2						
H3						

	1/944	1/1030	1/1030
	12	12	12
	30 - 90	30 - 90	30 - 90

Total Units	No.	Average Weight	% Unit	Total Units	No.	Average Weight	% Unit	Total Units	No.	Average Weight	% Unit	Total Units

Ground nesting birds

animal impact.

A. Estimated Total ADs (Livestock/Wildlife) _____
B. Estimated Days of Non Growth _____
C. Days of Bulk Feeding _____
D. Days of Drought Reserve Required _____
E. Total Days Grazing Required _____
F. Estimated Carrying Capacity _____

SUMMARY LIVESTOCK AND LAND PERFORMANCE
Calving/Lambing/Kidding _____ %
Avg Weaning Weight _____ Age _____ (Mths)
Daily Weight Gains _____ _____ Lbs
(Growing) (Nongrowing)

ADA/inch or ADH/mm rainfall _____
Total Yield (sold) per acre/hectare _____ Lbs

Figure 2-22. *Continued*

Sample Grazing Chart (Steps 8 - 13)

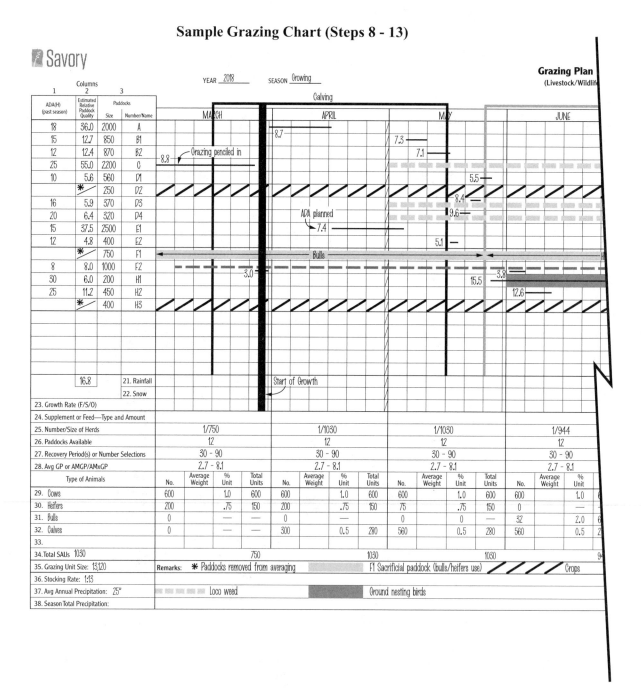

Figure 2-23. Because this is an open-ended plan, grazings have only been plotted through mid-June—the middle of the growing season in this case. The grazings for the remainder of the season will be planned as the plan is implemented. Paddocks D2 and H3 have been removed from the paddock average, since they are cropfields; and paddock F1 has been removed from the paddock average because it is a sacrificial paddock being used for bulls, then heifers, then bulls again on continuous graze.

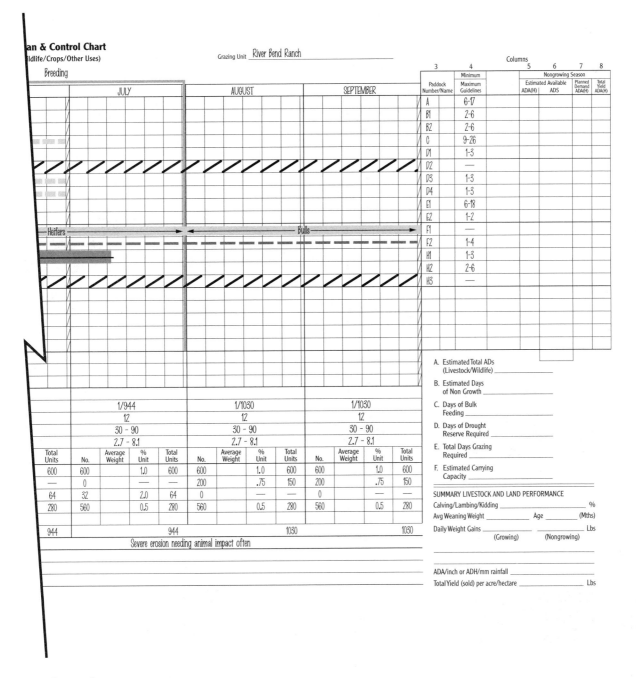

an & Control Chart
(ldlife/Crops/Other Uses)

Grazing Unit __River Bend Ranch__

Breeding

	JULY	AUGUST	SEPTEMBER

Heifers → ← Bulls →

	3 Paddock Number/Name	4 Minimum Maximum Guidelines	5 ADA(H)	6 ADS	7 Planned Demand ADA(H)	8 Total Yield ADA(H)
	A	6-17				
	B1	2-6				
	B2	2-6				
	C	9-26				
	D1	1-3				
	D2	—				
	D3	1-3				
	D4	1-3				
	E1	6-18				
	E2	1-2				
	F1	—				
	F2	1-4				
	H1	1-3				
	H2	2-6				
	H3					

Columns — Nongrowing Season — Estimated Available

JULY	AUGUST	SEPTEMBER
1/944	1/1030	1/1030
12	12	12
30 – 90	30 – 90	30 – 90
2.7 – 8.1	2.7 – 8.1	2.7 – 8.1

Total Units	No.	Average Weight	% Unit	Total Units	No.	Average Weight	% Unit	Total Units	No.	Average Weight	% Unit	Total Units
600	600		1.0	600	600		1.0	600	600		1.0	600
—	0	—	—	200	.75	150	200		.75	150		
64	32	2.0	64	0	—	—	0		—	—		
280	560	0.5	280	560	0.5	280	560		0.5	280		
944		944		1030		1030						

Severe erosion needing animal impact often

A. Estimated Total ADs (Livestock/Wildlife) _____

B. Estimated Days of Non Growth _____

C. Days of Bulk Feeding _____

D. Days of Drought Reserve Required _____

E. Total Days Grazing Required _____

F. Estimated Carrying Capacity _____

SUMMARY LIVESTOCK AND LAND PERFORMANCE

Calving/Lambing/Kidding _____ %
Avg Weaning Weight _____ Age _____ (Mths)
Daily Weight Gains _____ _____ Lbs
(Growing) (Nongrowing)

ADA/inch or ADH/mm rainfall _____
Total Yield (sold) per acre/hectare _____ Lbs

Figure 2-23. *Continued*

Recording SAUs

Type of Animals	No.	Average Weight	% Unit	Total Units
29. Cows/heifers	50/10		1.0/.75	57
30. Bulls	2		2.0	4
31. Ewes	300		5:1	60
32. Lambs	360		10:1	36
33. Rams	8		5:1	2
34. Total SAUs	165			159

← Highest (peak) from all months

Figure 2-24. The peak SAU figure, shown in the first column of row 34, is used to calculate the stocking rate (grazing unit size divided by peak SAU).

Rating Relative Paddock Quality (Growing Season Plan)

| | Columns | | |
| 1 | 2 | 3 | |
Actual ADA(H) (past season)	Estimated Relative Paddock Quality	Paddocks Size	Number/Name
18	36.0	2,000	A
15	12.8	850	B1
14	12.2	870	B2
25	55.0	2,200	C
10	5.6	560	D1
15		250	D2
16	5.9	370	D3
20	6.4	320	D4
15	37.5	2,500	E1
12	4.8	400	E2
11		750	F1
8	8.0	1,000	F2
30	6.0	200	H1
25	11.2	450	H2
13		400	H3

20 ADA x 320 AC = 6,400
6,400 ÷ 1,000 = 6.4

Average Paddock Rating

| 16.8 | 21. Rainfall |
| | 22. Snow |

Figure 2-25. The average rating for all paddocks shown in the box at the bottom of column 2, is derived by adding all figures in column 2 and dividing the sum by the total number of paddocks. Note that paddock F1 is blank because it was used as a sacrificial paddock in the last nongrowing season. If it is going to come into this growing season plan, then use the most recent ADA(H) rating you have for this paddock.

Figuring Grazing Periods (Growing Season Plan)

| 2 | 3 | | 4 | |
Estimated Relative Paddock Quality*	Paddocks SIZE	NUMBER/NAME	Minimum Maximum Guidelines	
36	2,000	A	6 – 17	
12.8	850	B1	2 – 6	
12.2	870	B2	2 – 6	
55	2,200	C	9 – 26	
5.6	560	D1	1 – 3	
3.8	250	D2		
5.9	370	D3	1 – 3	
6.4	320	D4	6 – 18 Days	
37.5	2,500	E1	1 – 2	
4.8	400	E2		
8.3	750	F1	1 – 4	
8	1,000	F2	1 – 3	
6	200	H1	2 – 6	
11.2	450	H2		
5.2	400	H3		
16.8		21. Rainfall		

*ADs in thousands; bottom figure (16.8) is average ADs

| 28 AMGP / AMxGP | | 2.7 – 8.1 | 2.7 – 8.1 |

Minimum recovery periods in the highest rated paddocks will be

Paddock	Recovery
A	33 – 6 = 27 Days
C	33 – 9 = 24 Days
E$_1$	33 – 6 = 27 Days

If the 6 days lacking in paddock C are distributed among any of the paddocks that have sufficient recovery time, all paddocks will have 30 days or more.

$$\frac{37.5}{16.8} \times 2.7 = 6.0$$
$$\times 8.1 = 18.0$$

Total of min. GPs = 33

Figure 2-26. Check to see that recovery periods are adequate in longer-grazed paddocks by adding together all the minimum grazing periods and subtracting the longest minimum grazing period from the total. If any recovery period is too short, add days to the minimum grazing periods in paddocks that can absorb them.

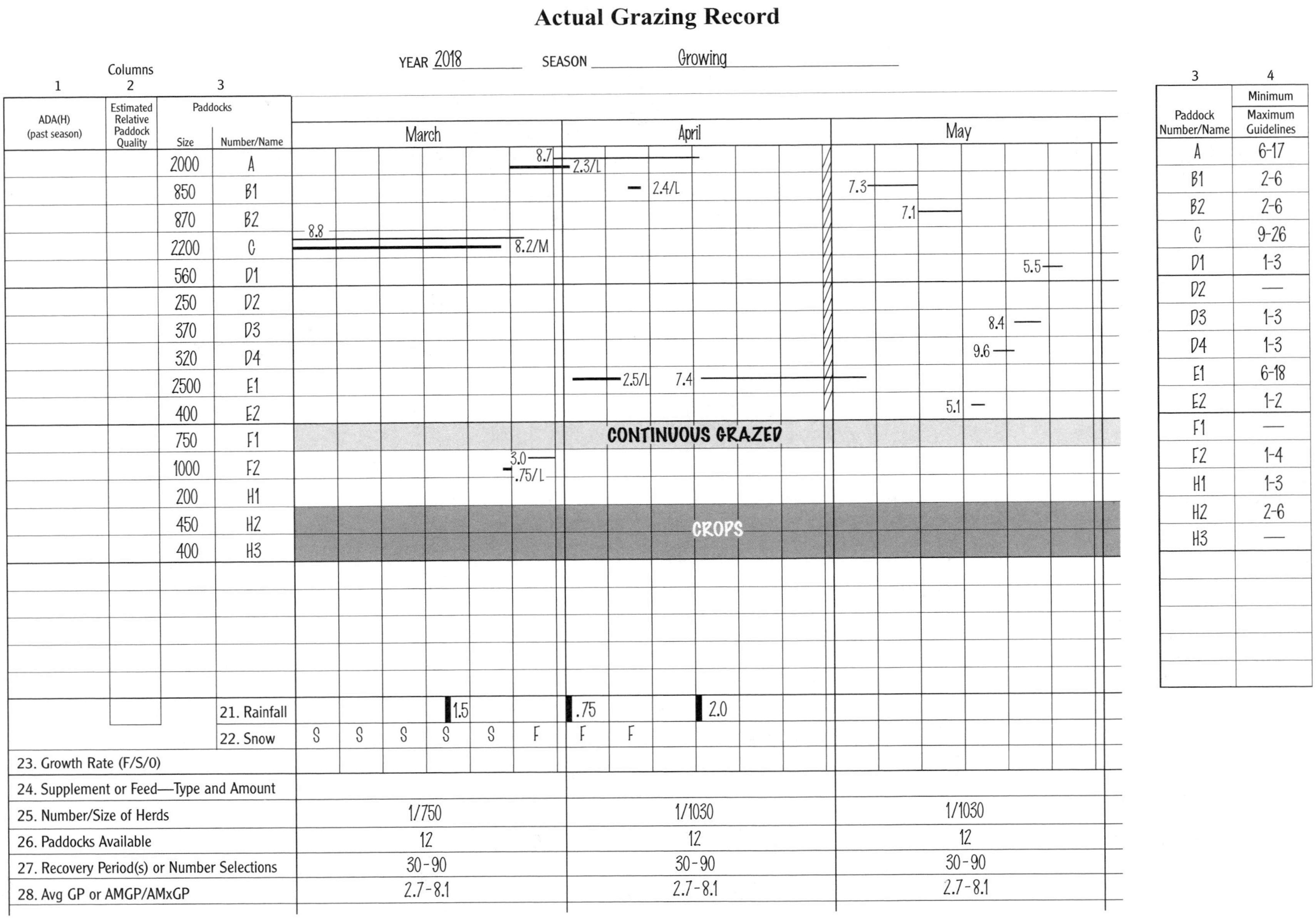

Actual Grazing Record

YEAR 2018 _____ SEASON _____ Growing _____

Paddock Number/Name	Minimum Maximum Guidelines
A	6-17
B1	2-6
B2	2-6
C	9-26
D1	1-3
D2	—
D3	1-3
D4	1-3
E1	6-18
E2	1-2
F1	—
F2	1-4
H1	1-3
H2	2-6
H3	—

ADA(H) (past season)	Estimated Relative Paddock Quality	Size	Paddocks Number/Name
		2000	A
		850	B1
		870	B2
		2200	C
		560	D1
		250	D2
		370	D3
		320	D4
		2500	E1
		400	E2
		750	F1
		1000	F2
		200	H1
		450	H2
		400	H3

Column headings: 1 Columns, 2, 3 Paddocks

Chart values (by paddock, March–May): A 8.8, 8.7; B1 2.3/L, 2.4/L, 7.3; B2 7.1; C 8.2/M; D1 5.5; D3 8.4; D4 9.6; E1 2.5/L, 7.4; E2 5.1; F1 3.0, .75/L (CONTINUOUS GRAZED); F2 1.5, .75, 2.0; H2 CROPS

	March	April	May
21. Rainfall	S S S	S F F	
22. Snow	S	S F F	
23. Growth Rate (F/S/0)			
24. Supplement or Feed—Type and Amount			
25. Number/Size of Herds	1/750	1/1030	1/1030
26. Paddocks Available	12	12	12
27. Recovery Period(s) or Number Selections	30-90	30-90	30-90
28. Avg GP or AMGP/AMxGP	2.7-8.1	2.7-8.1	2.7-8.1

Figure 2-27. Note that the inked-in grazings (heavier lines) do not match the grazings originally planned. After rain was received on March 18, growth became rapid a little over a week later (not immediately!), and the grazings dropped to the minimum time (i.e., from three days to one day). The grazing was light (L); 3.0 ADA was planned to be taken, but only .75 was actually taken. As soon as growth slows down, all penciled-in grazings would be replanned because the faster moves will have changed the plan.

Sample Grazing Chart (Steps 1 - 5)

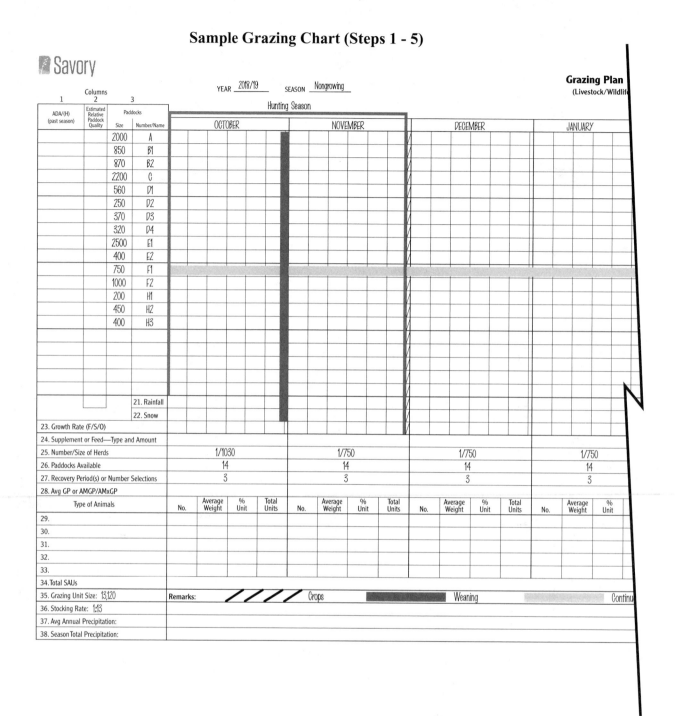

Figure 2-28. This partially completed chart shows major management concerns for this ranch in the nongrowing season and drought reserve (which begins in March)—hunting during October and November and the beginning of calving in March. The only special concerns noted in individual paddocks is the crop preparation that will begin in mid-March. Line 25 shows one herd each month, but that herd size changes in some months. Line 27 shows that there will be three selections taken over the nongrowing period and drought reserve.

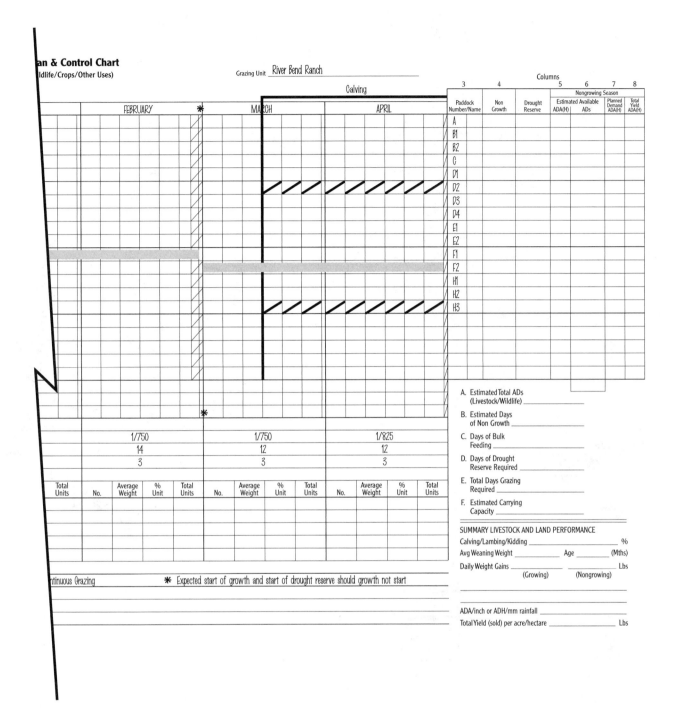

an & Control Chart
(Idlife/Crops/Other Uses)

Grazing Unit River Bend Ranch

Calving

FEBRUARY	✱	MARCH	APRIL	

Columns

3	4		5	6	7	8
			Nongrowing Season			
Paddock Number/Name	Non Growth	Drought Reserve	Estimated Available		Planned Demand ADA(H)	Total Yield ADA(H)
			ADA(H)	ADs		

Paddocks: A, B1, B2, C, D1, D2, D3, D4, E1, E2, F1, F2, H1, H2, H3

A. Estimated Total ADs (Livestock/Wildlife) _____

B. Estimated Days of Non Growth _____

C. Days of Bulk Feeding _____

D. Days of Drought Reserve Required _____

E. Total Days Grazing Required _____

F. Estimated Carrying Capacity _____

SUMMARY LIVESTOCK AND LAND PERFORMANCE

Calving/Lambing/Kidding _____ %

Avg Weaning Weight _____ Age _____ (Mths)

Daily Weight Gains _____ _____ Lbs
(Growing) (Nongrowing)

_____

_____

ADA/inch or ADH/mm rainfall _____

Total Yield (sold) per acre/hectare _____ Lbs

	1/750				1/750				1/825		
	14				12				12		
	3				3				3		

Total Units	No.	Average Weight	% Unit	Total Units	No.	Average Weight	% Unit	Total Units	No.	Average Weight	% Unit	Total Units

ntinuous Grazing _____   ✱ Expected start of growth and start of drought reserve should growth not start _____

Figure 2-28. *Continued*

Sample Grazing Chart (Steps 6 - 12)

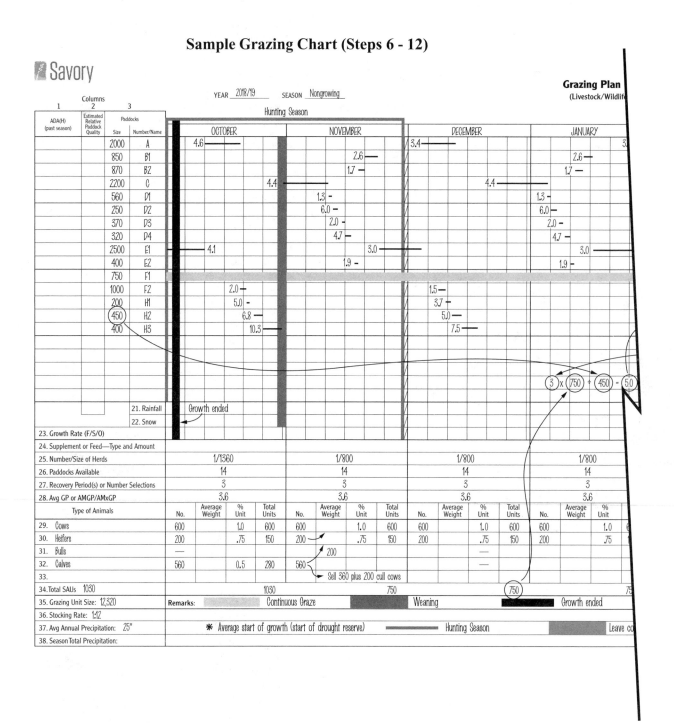

Figure 2-29. The drought reserve period covers March and April in this case; the cropfields (D2 and H3) will be grazed during the nongrowth period, but not in the drought reserve period as they will require cultivation and planting; the sacrificial paddock (F1) changes to F2 over the drought reserve period and will carry into the new growing season when it starts (which could be in March or April).

an & Control Chart
(ldlife/Crops/Other Uses)

Grazing Unit __River Bend Ranch__

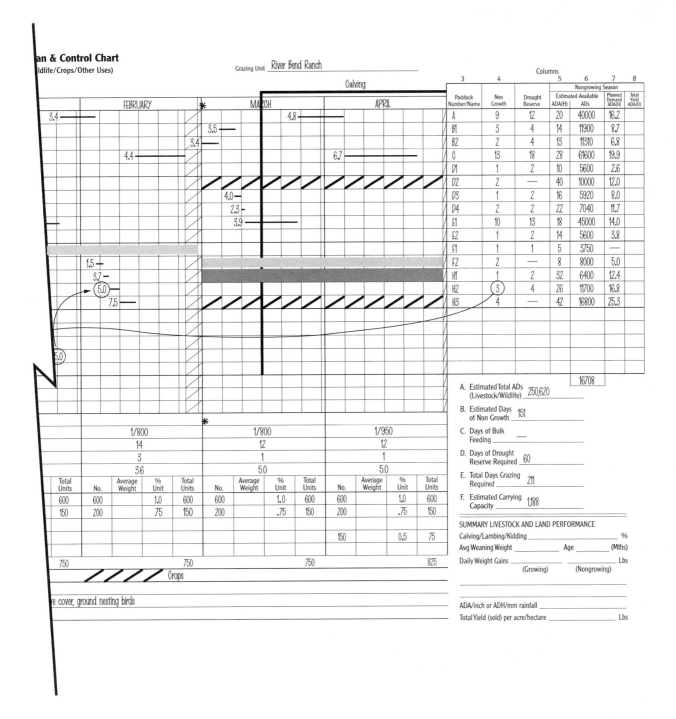

Calving

	FEBRUARY	*	MARCH	APRIL
3.4				4.8
		3.5		
		3.4		
	4.4			6.7

4.0
2.3
3.9

1.5
3.7
(5.0)
7.5

(5.0)

	3	4	Columns	5	6	7	8
				Nongrowing Season			
Paddock Number/Name	Non Growth	Drought Reserve	Estimated Available		Planned Demand ADA(H)	Total Yield ADA(H)	
			ADA(H)	ADs			
A	9	12	20	40000	16.2		
B1	3	4	14	11900	8.7		
B2	2	4	13	11310	6.8		
C	13	18	28	61600	19.9		
D1	1	2	10	5600	2.6		
D2	2	—	40	10000	12.0		
D3	1	2	16	5920	8.0		
D4	2	2	22	7040	11.7		
E1	10	13	18	45000	14.0		
E2	1	2	14	5600	3.8		
F1	1	1	5	3750	—		
F2	2	—	8	8000	5.0		
H1	1	2	32	6400	12.4		
H2	(3)	4	26	11700	16.8		
H3	4	—	42	16800	25.3		

16708

A. Estimated Total ADs (Livestock/Wildlife) __250,620__

B. Estimated Days of Non Growth __151__

C. Days of Bulk Feeding __—__

D. Days of Drought Reserve Required __60__

E. Total Days Grazing Required __211__

F. Estimated Carrying Capacity __1,188__

1/800	1/800	1/950
14	12	12
3	1	1
3.6	5.0	5.0

Total Units	No.	Average Weight	% Unit	Total Units	No.	Average Weight	% Unit	Total Units	No.	Average Weight	% Unit	Total Units
600	600		1.0	600	600		1.0	600	600		1.0	600
150	200		75	150	200		.75	150	200		.75	150
									150		0.5	75
750				750				750				825

Crops

SUMMARY LIVESTOCK AND LAND PERFORMANCE

Calving/Lambing/Kidding _____ %

Avg Weaning Weight _____ Age _____ (Mths)

Daily Weight Gains _____ _____ Lbs
                    (Growing)   (Nongrowing)

_____

ADA/inch or ADH/mm rainfall _____

Total Yield (sold) per acre/hectare _____ Lbs

ve cover, ground nesting birds

Figure 2-29. *Continued*

Rating Relative Paddock Quality (Closed Plan)

Columns

1	2	3			5	6	7	8
ADA(H) (past season)	Estimated Relative Paddock Quality	Paddocks			Nongrowing Season			
		Size	Number/Name		Estimated Available		Planned Demand ADA(H)	Total Yield ADA(H)
					ADA(H)	ADs		
		2000	A		20	40,000		
		850	B1		14	11,900		
		870	B2		13	11,310		
		2200	C		28	61,600		
		560	D1		10	5,600		
		250	D2	250 x 40 = 10,000	40	10,000		
		370	D3		16	5,920		
		320	D4		22	7,040		
		2500	E1		18	45,000		
		400	E2		14	5,600		
		750	F1		5	3,750		
		1000	F2		8	8,000		
		200	H1		32	6,400		
		450	H2		26	11,700		
		400	H3		42	16,800		

21. Rainfall

22. Snow

Average ADs → 16,708

Figure 2-30. The average animal-days (ADs) of forage for all paddocks, shown in the box at the bottom of column 6, is derived by dividing the figure in row A (lower right of chart, not shown here) by the total number of paddocks.

Summary Livestock Results

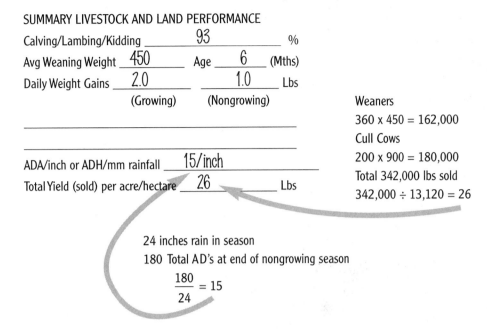

SUMMARY LIVESTOCK AND LAND PERFORMANCE

Calving/Lambing/Kidding _____93_____ %

Avg Weaning Weight __450__ Age __6__ (Mths)

Daily Weight Gains __2.0__   __1.0__ Lbs
         (Growing)   (Nongrowing)

_____

_____

ADA/inch or ADH/mm rainfall __15/inch__

Total Yield (sold) per acre/hectare __26__ Lbs

Weaners
360 x 450 = 162,000
Cull Cows
200 x 900 = 180,000
Total 342,000 lbs sold
342,000 ÷ 13,120 = 26

24 inches rain in season
180 Total AD's at end of nongrowing season

$$\frac{180}{24} = 15$$

Figure 2-31. This is also a summary of land performance, an important measure of which is the total yield per acre or hectare of products sold.

HOLISTIC ECOLOGICAL MONITORING

PART 3

HOLISTIC ECOLOGICAL MONITORING
Keeping Management Proactive

A FUNDAMENTAL PRINCIPLE OF HOLISTIC MANAGEMENT is that success requires planning—and successful planning requires implementation (of course), monitoring, controlling, and replanning. *Holistic Management* makes this point in many ways and contexts, and earlier chapters in this handbook have repeated it. In *any* situation you manage, you should be monitoring to make happen what you want to happen—to bring about desired changes in line with your holistic context. Monitoring developments in the ecological sphere, however, deserves its own treatment because much of what we do as land managers may lead to unanticipated effects. Anytime you plan to alter ecosystem processes in any way, *assume you could be wrong* because the land is more complex than humans will ever understand.

The livestock industry traditionally monitors many aspects of animal performance. In the cattle business, the statistics on conception rates, bull performance, daily gain, weight per day of age, calf and weaner weights, more or less define the quality of an operation in traditional terms. With the development of Holistic Management, ranchers are beginning to understand what crop farmers have long understood: that, in general, *yield per acre or hectare is more important to profit than yield per plant or animal.* While it is important to monitor animal performance, it is even more important to monitor the land's performance, its ability to convert sunlight to grass (and other plants) and thus to saleable livestock products, as well as to other potential enterprises, such as wildlife, recreation and ecosystem services, and, ultimately, money.

In assuming that any action we plan to take could go wrong because of the land's complexity, we want to have the earliest possible warning so we can make changes before damage is done. For instance, a drop in conception rates shows a problem—but after the fact and without any clue of how to correct it. Many of the numbers we ardently compile and ponder fail us in the same way.

Obviously, you can better steer a ship looking over the bow rather than back at the wake, but only if you know what to monitor. Ideally, ecological monitoring should pick up changing conditions and deviations from plan

129

as early as possible so you don't miss an opportunity to change course and replan. As you apply any one of the management tools—fire, technology, rest, or living organisms (including animal impact and grazing)—you will need to determine what criteria you can monitor that will give the earliest warnings of adverse change. Monitoring changes in plant or animal species, a common practice, is a measurement that comes too late, indicating considerable change has already occurred that may not have been in line with your holistic context. You want to detect changes well before that point whenever possible.

You must address this challenge on three levels. First, cultivate a general and ongoing awareness of the condition of the four ecosystem processes (water cycle, mineral cycle, energy flow, and community dynamics) and how the tools you apply affect them.

Second, each year carry out an annual assessment of the soil surface and the life upon it based on one of the procedures described in the second section of part 3, "Monitoring Your Land," which will help you to predict changes and trends.

Third, if you are managing livestock, you must monitor plant growth rates, water supplies, the development of unfavorable grazing patterns, and so on, when working to a holistic grazing plan.

These three levels of monitoring complement each other. When practiced together they will quickly reveal the dynamism present in all landscapes. Most people who make a living from the land or love it deeply find this revelation fascinating. The resulting habit of observation becomes addictive but hard to communicate to people who have not tried it. A Colorado rancher once confessed to lying to his guests to avoid the trouble. "At chore time I tell them I'm going to check the cattle," he said. "They wouldn't believe that I'm really checking the grass."

Like the two preceding parts of this book, part 3 presents some basic concepts and examples that illustrate aspects of Holistic Management, in this case ecological monitoring, and then a guide to a formal monitoring procedure, of which there are two in this chapter: basic and comprehensive. These procedures specifically address the concerns of those managing brittle-environment rangelands or grasslands, though there are notes for those managing grasslands much lower on the brittleness scale. They assume that grazing and browsing animals—domestic and wild—are present for at least part of the year. Those who routinely till, replant, and fertilize their pastures are in effect managing croplands, and these procedures won't apply. But anyone managing croplands should consider using a good soil-testing laboratory to monitor the life within their soils. Good resources include Neal Kinsey (http://www.kinseyag.com) and Elaine Ingham (http://www.soilfoodweb.com) in the United States, and Christine Jones (http://amazingcarbon.com) in Australia.

If you are managing wetlands or forest, these same procedures can be used to monitor conditions in the rainfall catchment areas (or watersheds) leading into a wetland or surrounding a forest. Within a forest, and particularly when livestock are used to help control the buildup of combustible understory, these procedures will provide essential information, but they will need to be augmented by more sophisticated measurements of the overstory. Using these procedures may be very helpful on grazed wetlands or wetlands that accumulate excessive standing litter and lose biodiversity when overrested.

The advice contained here, of course, is not exhaustive, as the possibilities have no end. It nevertheless provides a starting place. And if you haven't looked at land this way before, no land will ever appear the same again.

Mastering the Basics

AS MENTIONED EARLIER, good monitoring depends on a broad awareness of the state of the land, but this means training yourself to look for a broad range of specific things. If you happen to be a hunter, you know that putting your mind on deer tends to program your eyes to see deer. Tomorrow you might walk the same woods hunting rabbits or turkeys and not notice the buck that would have stood out clearly yesterday. With Holistic Ecological Monitoring, you must consciously direct your thoughts in order to see. And don't just think about the ground. Look in books and records, in the expertise of others, and in your own experience.

Historical Data—Where You've Been

Monitoring has no meaning except in relation to your goals and your plans for reaching them. Progress toward any goal, however, implies a starting point—and your monitoring will be much more perceptive and useful if you know where that starting point lies in relation to the past use and potential use of your land. If you have read, for instance, that pioneers a hundred years ago harvested winter hay from what has since given way to desert, this knowledge will allow you to assess the present community relative to its potential, and alert you to vestiges of lost plant species that your plans could revive. If you've just purchased a tree farm in the Carolinas and don't know that it once supported a prosperous tobacco plantation, you might suppose that it could never grow anything but loblolly pine and therefore not look for improvement.

When you begin managing a piece of land, photograph the characteristic sites, take notes of their condition, and try to reconstruct what has happened there in the past. Useful information might include the location of springs and streams that have dried up or become intermittent, the dates and extent of past fires and floods, and observations about changes in the types of crops that can be grown, the species of wildlife, and the makeup of plant communities. You will want to back up these general observations with hard evidence from fixed-point photographs and data from your routine monitoring— but again, general observations are equally important. Your notes might look something like those for the Sheep Creek Drainage (see box 3-1).

Creating Your Future Landscape—Where You Want to Go

In your holistic context you should have described the four ecosystem processes, as they will have to be functioning—far into the future—if the land is to sustain the quality of life you envision. Progress toward the future condition of these four processes is chiefly what you will be measuring in your annual monitoring. You will also be looking for signs of progress in creating the landscape features you may have mapped if your land is extensive and the terrain varied: a tree-lined riparian community, wetlands, dense brush areas, or open grassland communities.

Once you have monitored your land for the first time, it's a good idea to summarize the status of the four eco-

Box 3-1. Historical Data: Sheep Creek Drainage, June 2018

According to county historians, Sheep Spring was an important watering point on sheep drives to the railhead at Upstart. Over 10,000 sheep passed through in the summer of 1879—evidence in graffiti on rocks. The creek must have been quite different as accounts mention frequent fights between herders when bands of sheep mingled across a shallow river that meandered through a meadow.

Tradition says Indians had an antelope trap in the canyon at the south end of the valley. Arrowheads are found there still.

In 1903, there was a grass fire hot enough to destroy a prospector's cabin.

The Vee Bar, a huge open-range cattle operation, included the valley in its summer range until 1929. Intermittent use by squatters continued till 1940, when it was fenced. Continuous grazing continued until 2000, when it was acquired by Vesuvius Life Insurance Co. No stock or formal management until 2010.

Box 3-2. Current Landscape Description: Sheep Creek Drainage

Community Dynamics

Community is definitely at a lower successional stage than in its glory days, consisting mostly of sage and annual grasses. There are many young piñons, and anthills are abundant. Sage, winterfat, cliff rose, and other brush show signs of severe past overbrowsing followed by regrowth that must have begun in 2000. Some winter use by deer is evident. There are isolated examples of perennial grasses (including sand dropseed, giant dropseed, Indian rice grass, western wheatgrass) often associated with yucca or less frequently with sage, as well as some mats of blue grama and a fair amount of spiny muhly.

Water Cycle

The water cycle is generally ineffective and evidenced by a large percentage of bare ground. Most perennial plants are on pedestals, often with exposed roots. Soil capping is pervasive, often mature and black with algae. There are many rills and small gullies, and most litter is washed into heaps and banks, showing the force of runoff. Sheep Creek is an intermittent arroyo [stream] with six- to ten-foot (two- to three-meter) vertical banks broken occasionally by old stock and game trails.

Mineral Cycle

The mineral cycle is also generally ineffective. Dung pats three years and older are evident everywhere. Perennial grasses show accumulations of past growth. Annual grasses are largely washed or blown away by midwinter. Deeper-rooted plants, sage, and piñon are thriving.

Energy Flow

The energy flow is obviously not high because of large amounts of bare ground. Most of the sunlight energy converted to forage is not harvested by animals.

system processes, so you have a baseline, or starting point, that you can compare against in subsequent years. How the manager in the Sheep Creek example might do that summary is described in box 3-2.

The Photo Record

Photographs show changes on the land better and more dramatically than any other record. Later we'll see how to take a measurement of your land's health, including several photos taken from the same spot each year. Snapshots taken of the same scene in different seasons across a period of years broaden the record and may lead to important insights.

Carry a camera, even if it's just in your smartphone, and get in the habit of shooting interesting anomalies, and reshooting the same scene for comparison. In this way you can monitor the healing of a gully, the development of a fence-line contrast, animal impact around a water point or gate, the effect of the herd in one corner of a paddock during a heavy rainstorm (and the same spot in the fol-

lowing season), the formation of trails, and any number of other concerns.

Smartphone cameras (if they have good lenses) are an excellent choice because you tend to have them with you most of the time. And, as with any digital camera, you can check the quality of your photos immediately after taking them and reshoot the photos if necessary. Storing the photos on a computer or spare hard drive is also easy with the help of software programs, many of them free, designed to simplify photo identification and filing. Just make sure that you set up a system at the outset so you can always identify and retrieve the photos you need.

The Soil Surface Is Key

The earliest changes on any piece of land are most likely to occur at or near the soil surface, but particularly so on land that has been reduced to a high percentage of bare ground. They could show up as a change in plant spacing, soil litter cover, soil density, soil aeration, porosity, organic content, insect activity, seedling success, quality of water runoff, and a host of other things.

Depending on what your monitoring indicates, you will either continue to apply the tools as you have been or you will need to make adjustments. Obviously, if all is going as planned when the particular tool was selected, no adjustment is necessary. If not, you will have to diagnose what went wrong and develop alternatives you can use as you replan.

The main thing to remember is that you are looking for basic information you can photograph, measure, and understand, that indicates *to you* what changes are taking place, rather than a mass of data that is of little practical use.

The Brittleness Factor

All terrestrial environments, regardless of total rainfall, fall somewhere along the continuum from nonbrittle to very brittle. For simplicity, we refer to this continuum as a 10-point scale—1 being nonbrittle and 10 being very brittle.

Environments can be classified along this continuum according to how well humidity is distributed throughout the year, and how quickly dead vegetation breaks down (rapidly through biological decay, or slowly through chemical oxidation and weathering). At either end of the scale and along it, environments are likely to respond similarly to the different management tools—fire, for example, will tend to produce bare soil and rapid oxidation of plant material; herbicides will tend to kill plants; high ani-

mal impact will always speed the return of plant material to the soil.

However, the tool of rest produces profoundly different effects across the brittleness scale. Rest can be applied in two forms. *Total rest* refers to the withholding of any form of disturbance (fire, grazing, machinery, etc.) for a considerable length of time. *Partial rest* occurs when grazing animals are present but behave so calmly in the absence of pack-hunting predators that a large proportion of the plant life and soil surface remains undisturbed despite their presence.

In very brittle environments, either form of rest will lead to an increase in bare ground. The exception is high-rainfall brittle environments (generally in the tropics) where rest tends to damage or destroy grassland and move the vegetation to woodland with enough leaf fall to provide permanent soil cover.

As any land approaches the nonbrittle end of the scale, rest in either form becomes increasingly powerful in restoring full health and biological diversity to communities. At the lowest level (1), in tropical wet forests, for example, resting the land restores even the most damaged environments, and that is why the ruins of past failed civilizations are found under fully recovered forest. A little higher up the brittleness scale, while land may never exhibit a high percentage of bare soil between plants, it is common to find that, when rested excessively, much of the soil surface is dominated by moss and lichen communities. The ground will almost never look bare. Overgrazing of plants may produce short annual runner grass swards, and overtrampling a light covering of moss, both of which provide little nourishment to grazers. But partial or total rest can produce thick layers of moss and, particularly in northern climes, lichen, that have even less forage value. Knowing approximately where you are on the brittleness scale is essential when interpreting the results of your monitoring.

The distribution of the precipitation, as well as the elevation, temperature, and prevailing winds, clearly affects the day-to-day distribution of humidity, and this links very closely to the degree of brittleness. The poorer the distribution of humidity, particularly in the growing season, the more brittle the area tends to be, even though total rainfall may be high. Very brittle environments commonly have a long period of nongrowth that can be very arid.

SIDEBAR 3-1

Nonbrittle Environments

Completely nonbrittle environments are characterized by the following:

- Reliable precipitation regardless of volume
- Good distribution of humidity throughout the year as a whole
- A high rate of biological decay in dead plant material, which is most rapid close to the soil surface (thus dead trees rot at their bases and topple over relatively quickly)
- Speedy development of new communities on any bare surface
- The development of complex and stable communities, even where they are not physically disturbed for many years

In such environments it is virtually impossible to produce or maintain millions of acres or hectares where the ground between plants is bare, other than on croplands that are continually exposed by machinery.

SIDEBAR 3-2

Brittle Environments

Very brittle environments are characterized by the following:

- Unreliable precipitation, regardless of volume
- Poor distribution of humidity through the year as a whole
- In the absence of sufficient grazing animals, chemical (oxidation) and physical (weathering) breakdown of dead plant material, generally slow and from the upper parts of plants downward (thus dead trees remain standing for many years; a dead perennial grass clump can stand for several decades)
- Very slow development of communities from bare soil surfaces unless periodically disturbed (commonly by bunched, herding animals)
- Soil surfaces that can be covered with algae and lichens for centuries unless adequately disturbed

In such environments it is very easy to produce millions of acres or hectares where the ground between plants is bare or capped with algae and lichen, merely by resting the land excessively (through partial or total rest), burning it frequently, or overgrazing many grass plants. Such areas tend to maintain biodiversity and stability only when adequately disturbed by herding animals that move sufficiently to neither overgraze nor overbrowse.

If you were dealing with land at 7 or 8 on a scale of 1–10, it would tend to have most of the features of the very brittle extreme. Land at 2 or 3 would tend to have most of the features of the nonbrittle extreme. The less brittle environments of the middle range (4 to 6) may be more difficult to identify, but precision is not necessary because the land is more forgiving of errors in judgment.

Key Land Health Indicators: The Four Ecosystem Processes

In your holistic context you described the environment needed to sustain your quality of life (and that of future generations) in terms of the four ecosystem processes—all of which need to function well. Here are some guidelines to help you in gauging the health of each one of them:

Community Dynamics: Assessing Succession

Community complexity is the key indicator of the land's stability, resilience, productivity, and health. Manipulating the level or degree of complexity within a community is the principal way to achieve the landscape you desire, but you first need to identify whether a community is increasing or decreasing in complexity. This is always a subjective exercise, but it becomes easier if you think in terms of *succession*—the sequence of stages through which a biological community develops. Roughly speaking, bare ground gives way to algae/lichen/moss, grasslands, brushlands, and forest in a gradual, often staggered, buildup of plant and animal species diversity and biomass.

Older textbooks like to talk about "climax" communities as a definite end point of succession, but in reality few environments (especially brittle ones) enjoy stable conditions long enough to determine what this is. In many instances, brittle environment communities in which large grazing animals had been absent for some time were defined as climax communities, based on the plant species present. However, such a climax community would be very different after the three or four days it took thousands of bison to pass over it, and yet those animals were as much a part of the community as the soil and plants they consumed or trampled. Increasingly, scientists are coming around to the view that in brittle environments the fluctuations in vegetation following disturbance are normal and natural, and that these environments, when disturbed, are considerably more productive and generally more stable.

Thus you can judge the level of succession and its general direction by a variety of signs, some of which may conflict. Here are some of the main ones:

- **Simplicity versus complexity:** A wide diversity of species (rather than vast numbers of one species) suggests an advanced community. Even if total biological mass is impressive, monocultures of plants or vast numbers of one animal usually indicate a lower level of community development. Seasonal diversity is also important. In some parts of North America an absence of either cool-season or warm-season grasses, for example, suggests a simplified community.

- **Annual versus perennial:** A predominance of annual plants indicates a low level of community development (low succession). Among grass species, annuals generally grow a seed head on every stalk and die at the end of the growing season. Perennial grasses have many stalks that produce only leaves and will probably show some sign of the previous year's growth. Perennial grass roots usually display greater spread and depth and affect soil porosity more than annual grasses do. With rare exceptions, perennial plants contribute far more than annuals to soil cover and stability through their greater root depth and longevity. They also give a more consistent picture of community health, because huge numbers of annuals may burst forth only in seasons where the right conditions coincide.

- **Presence of certain plants and animals:** Generally, complexity of species is a more reliable index of community health than mere numbers, but most species thrive only at certain successional levels as illustrated in figure 3-1. Early occupants of any community tend to reproduce rapidly and depend on a simple environment. High production of seeds adapted to disperse

Organisms Thrive at Certain Successional Levels

Figure 3-1. There is no such thing as a hardy plant or animal. All organisms thrive at certain successional levels. Some, like humans and coyotes, range over a wide spectrum of successional levels. Others, like prairie dogs or flying squirrels, are confined. As succession advances, better mineral and water cycles will support a greater variety of species.

and penetrate capped soil is typical of the grasses occupying simple, or low-successional, communities. Many kinds of rodents and harvester ants are low successional. Moose and fir trees are typically high-successional organisms found in more complex communities. Among domestic animals, dairy cattle have a limited niche high in succession, while sheep and goats span a broader range of communities.

- **Status of youngest age class:** The direction in which the community is moving often shows first in the presence or absence of young plants or animals. If the young of perennial grasses or high-successional animals do not survive, that foretells decline—as would extraordinary reproductive success of low-successional organisms. Likewise, young animals or plants that increase diversity or fill high-successional niches indicate advance. Among grasses, a drop in successional level will at times show up as an increase in species that spread effectively by rhizomes or stolons rather than seeds.

- **Presence of woody plant species:** Woody forbs, brush, and trees usually reflect an advance in succession—as in the case of a tropical cornfield returning to jungle in a nonbrittle environment, or a pasture returning to forest in a less brittle environment. In brittle environments this matter is more complex, and woody species often flourish in declining grassland. This is not an advance when it represents a loss of diversity, and it may be a passing phase as damaged water cycles eventually kill off the woody plants too. In many instances, you will want to hold the community at a particular level and not allow too far an advance; for example, grassland not being allowed to move to woodland in a medium- to high-rainfall brittle environment, or any nonbrittle environment.

- **Status of ground cover:** Since bare, capped ground is by definition the bottom end of succession, reflecting a very impoverished community, the earliest sign of change is often

an increase or decrease in the space between perennial plants—particularly grasses. Many organisms will not propagate without open friable soil. Areas too arid to sustain a cover of living plants require litter. Changes in litter and capping often precede changes in succession as the complexity of a community decreases or increases. In many arid or semiarid grassland and savanna situations *litter, more than any other factor, determines which way the community will move because litter provides most of the soil cover.*

In less brittle environment grasslands, it may be hard to find bare ground, but there is often more than you'd expect if you get down on your hands and knees and look for it. In these areas, bare ground is more commonly found where plants have been overrested, or on poor soils that contain many overgrazed plants. In nonbrittle environments overrested soils may be capped with a thick layer of moss or lichens.

- **Remnants of old communities:** Remnants of plant species that once thrived in more complex, higher successional communities may hang on in sites protected by thorn bushes, cactus, rocks, and the like. They document the land's decline, indicate its potential, and provide the seeds for a comeback.

- **Economic uses:** When an area's livestock industry has to shift from cattle to sheep to goats or from production of meat to primarily hair, the community is generally declining.

Water Cycle: Assessing Its Effectiveness
The water cycle's effectiveness can be assessed by studying numerous factors, including erosion, permeability and effective rain, and plant habitat.

Erosion
Erosion from wind and water obviously indicates a noneffective water cycle. Beneficial use of water simply does not include moving soil. Obvious gullies indicate advanced

erosion. More pervasive, however, and equally destructive is sheet erosion, which leaves less obvious traces. Look for the following signs:

- **Pedestaling of plants and rocks:** Sheet erosion, either by wind or water, will cut away bare soil, leaving plants up on little platforms. Occasionally the plant loses the battle, and the soil is snatched from beneath it, leaving the roots exposed as stilts, as shown in figure 3-2. In some cases, plants actually accumulate soil, so the difference between the eroding surface and protected spots becomes extreme. In some parts of the American Southwest bushes like greasewood and Mormon tea now sit on mounds two or three feet (one meter) tall. Occasionally, plant pedestaling occurs because freezing soils expand, pushing plant bases upward and exposing roots. However, such pedestaling would be exacerbated by exposed soil, which results in greater temperature extremes at the surface.

- **Flow patterns on bare ground:** Moving water or wind scours the soil, leaving patterns, as shown in figure 3-3. Multiplied by the area of land involved, the tonnage of soil carried off is substantial. Equally important, it is difficult to accumulate organic matter under such conditions.

- **Litter banks:** Piles of litter jammed between plants and rock, which form small silt traps, as shown in figure 3-4, are a sign of more runoff of water than desirable, but the litter banks do stop some of it—showing once again the importance of litter.

- **Siltation in low points:** The silt dropped behind check dams, on gentle slopes below hillsides, or in streams, as shown in figure 3-5, means excessive and damaging runoff, even when it does not produce gullies.

- **Splash patterns:** Raindrops, especially big ones that fall from tall trees, dislodge and break up large particles of soil when they hit bare ground.

Figure 3-2. Plant pedestaling. This bunch grass plant is standing a good three inches (7.5 centimeters) above the soil, due to erosion. The exposed roots indicate how much soil has washed away.

Figure 3-3. Flow patterns. The movement of silt across the surface of this land is a result of severe sheet erosion.

Figure 3-4. Litter banks. Litter has caught between two plants creating a miniature dam. Silt has been deposited above the "dam," creating a flat spot, while below the dam, which is over two-inches (five centimeters) high, soil continues to wash away.

Figure 3-5. Siltation. The beginnings of a stream, which lies at the base of a large sheet-eroding hillside, has filled with silt.

This action can move a surprising amount of soil and badly degrade the soil surface. Check the height of the mud splatters on plants and fence posts.

- **Dunes:** Dunes are the end product of wind erosion, often forming far from the actual erosion site. Even small dunes cause problems when they move and smother or sandblast vegetation.

Permeability and Effective Rain

Ideally, precipitation should enter the soil where living organisms can use it. This of course depends on ground cover, soil type and condition, aeration, organic content, slope, and other factors. If you know how different sites respond to rain, you will be able to plan better use of sites and effective use of tools.

Long-term changes in soil permeability can cause dramatic changes in the water table and the disappearance or reappearance of springs. Hard, capped soil and lack of ground cover or litter indicate a problem with infiltration and aeration as well as excessive loss of water through runoff across the soil or from later soil surface evaporation.

You can get a good idea of the situation by simply pouring a quart (liter) or so of water on the ground and timing how long it takes to disappear. If you wait five minutes and probe a bit with a shovel, you can also find out how effectively it penetrates, as shown in figure 3-6. A similar check after a short rain will give you some idea of how much growth response to expect from that rain.

A Simple Water Cycle Test

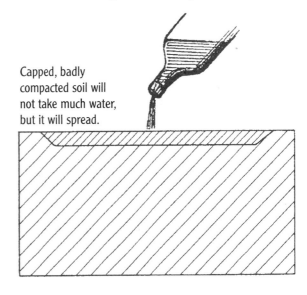

Capped, badly compacted soil will not take much water, but it will spread.

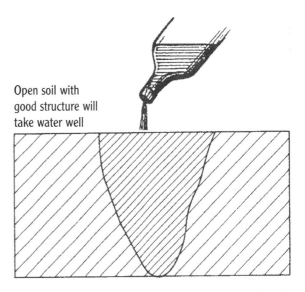

Open soil with good structure will take water well

Figure 3-6. Ideally, you want water to penetrate and soak into the soil, and that will depend on the nature of the soil surface and the soil type.

Plant Habitat

Plant types and associated animals will vary as soils range from dry to moist, from sealed to well aerated, from well drained to waterlogged, and as water tables rise or fall, as illustrated in figure 3-7. Changes in soil cover and community complexity will result in changes in the water cycle's effectiveness.

A good look at your own land will probably reveal which species you can expect to increase and which will decrease as the water cycle improves. In this con-

Grass Types from Habitats That Reflect Water Cycle

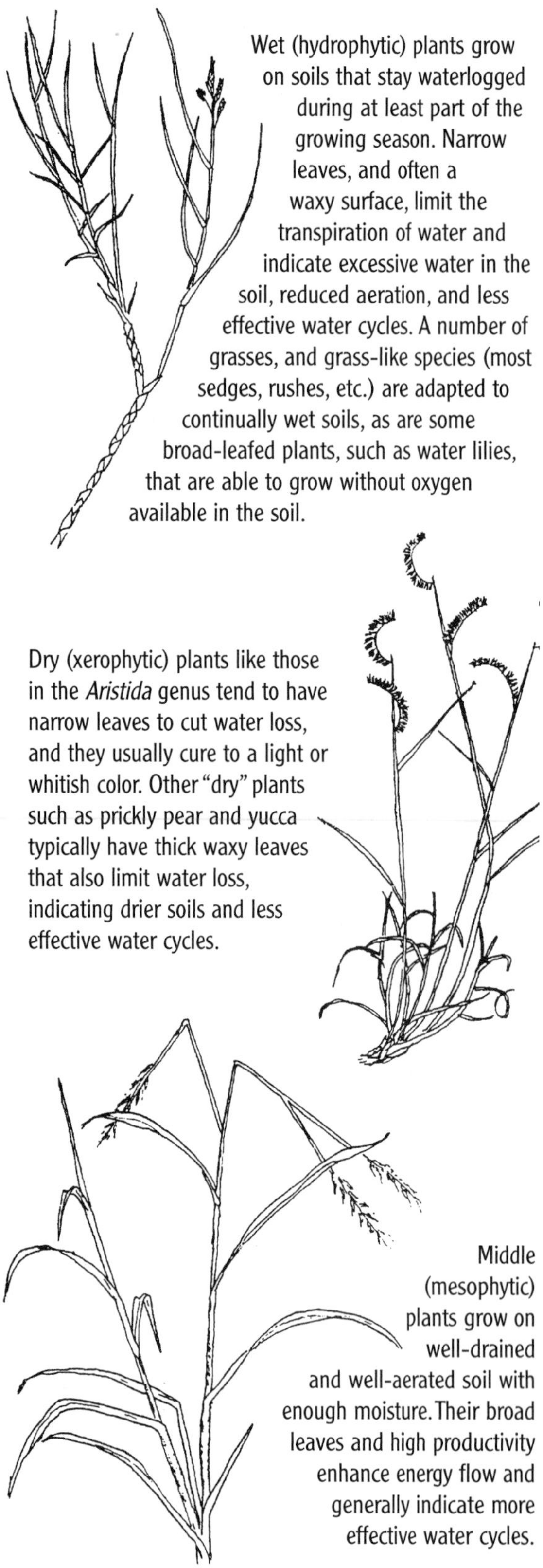

Wet (hydrophytic) plants grow on soils that stay waterlogged during at least part of the growing season. Narrow leaves, and often a waxy surface, limit the transpiration of water and indicate excessive water in the soil, reduced aeration, and less effective water cycles. A number of grasses, and grass-like species (most sedges, rushes, etc.) are adapted to continually wet soils, as are some broad-leafed plants, such as water lilies, that are able to grow without oxygen available in the soil.

Dry (xerophytic) plants like those in the *Aristida* genus tend to have narrow leaves to cut water loss, and they usually cure to a light or whitish color. Other "dry" plants such as prickly pear and yucca typically have thick waxy leaves that also limit water loss, indicating drier soils and less effective water cycles.

Middle (mesophytic) plants grow on well-drained and well-aerated soil with enough moisture. Their broad leaves and high productivity enhance energy flow and generally indicate more effective water cycles.

Figure 3-7. Plant types and associated animals will vary as soils range from dry to moist, from sealed to well aerated, from well drained to waterlogged.

text, remember that poor aeration, which occurs in both waterlogged and capped or compacted soils, affects plant growth as much as a lack of water because it decreases the effectiveness of the water cycle.

Mineral Cycle: Assessing Its Effectiveness

The mineral cycle, like community dynamics, has many aspects. Most of them, though, boil down to three questions you can check by observation:

1. Are minerals visibly cycling?
2. If not, what happens to them?
3. How well are corrective measures working?

The health of the mineral cycle shows in the following ways:

- **Breakdown of litter (especially dung):** A poor mineral cycle may allow dry dung pats to linger for years. In the more brittle environments, keep your eye on the amount of dead material that stays on plants and oxidizes, turning gray like an old fence post, instead of returning to the soil. In less brittle environments, look for distinct litter layers. If the soil has been managed well, it will be difficult to see where the litter stops and the soil begins. A new, undecayed layer of litter, which may be no more than bits of twigs, leaves, and dung on bare ground (what we call litter 1) can be seen in all environments. Deeper litter that is decaying and being incorporated into the soil (litter 2) is more easily seen in less brittle environments.

- **Activity of soil organisms:** Most of these are microbial and can't be seen, but worms, ants, and burrowing rodents all enhance the mineral cycle. When biodiversity is low some of these may boom to annoying numbers. The remedy often lies not in killing them directly but in increasing the diversity in the community (advancing succession) to such a degree that the offending species, while remaining in the community, no longer dominates but merely adds to the community's complexity.

- **Presence of plants of varying root depth:** Water transports many minerals from decomposing organic matter through soils. Deep-rooted plants can recapture and return leached minerals to the surface for growth; they can also access soil minerals that are available only from deeper parts of the soil. In high-rainfall areas, loss of mineral-binding organic matter allows leaching to become extreme. Thus farmers in wet climates must either apply extraordinary doses of fertilizer to "worn-out" pastures and croplands, which further damages soil life, or begin to regenerate them through better management of the ecosystem processes.

- **Livestock consumption of mineral supplements:** Livestock's appetite for supplements often reflects missing elements in the natural mineral cycle. Livestock exposed to free-choice, cafeteria-style minerals will select whatever they don't get from grazing. In the case of some trace minerals, they may add enough to the natural cycle through their dung to eventually reduce the need for supplements. As plant species concentrate minerals differently, a change in supplement consumption may also show the connection between the mineral cycle and biological diversity.

- **Deficiency symptoms in plants and animals:** Poor mineral cycles may produce rough coats, infertility, or other weaknesses in domestic or wild animals. In crop plants leaves may yellow, curl, or develop brown spots or other symptoms. Soil tests tell part of the story; however, minerals may be present but unavailable due to a variety of factors, such as a noneffective water cycle, a successional level too low to support microorganism populations that convert organic matter into humus and in the process make minerals available to the plant, inadequate concentrations of other minerals, or poor distribution and availability of minerals to plants.

Waterlogged soils or capped soils that are poorly aerated appear to "lock up" minerals that become available to plants once more as soil aeration improves. Finally, the ratio of minerals present for plant use may be so unbalanced that the plants are unable to adequately utilize certain ones.

- **Soil pH, sodium, and salination:** Extreme acidity or alkalinity makes many nutrients unavailable to plants. Salination occurs when damaged vegetation gives way to bare soil, through which water evaporates. As the water is drawn up through the soil profile, it carries salts with it that are left at the soil surface because evaporating water is fresh or "distilled." Excessive sodium destroys soil structure. More often than many like to admit these problems are greatly exacerbated by the lack of organic material that is characteristic of impoverished biological communities.

Rather than countering problems by merely testing the soil and adding the missing ingredient or removing the offending one, consider that covering soil (with plants or plant litter) and increasing complexity in the community, and thus advancing succession, might offer a better solution. Plants with varying root depths will recycle leached minerals, and a management plan that returns organic material to the soil will impede leaching and offset the effects of extreme acidity, alkalinity, sodium, and salination.

For insight into the mineral cycle, monitor community dynamics with an eye to the following:

- Diversity of plants, including nitrogen-fixing legumes
- Agents of decomposition, including the breakup of old plant material through animal impact
- The trend (increasing or decreasing) of soil cover and organic material on and in the soil
- Varying root depths

Use your observations of all these factors in management decisions. Holistic decision making may well show that composting, the manure spreader, more paddocks, increased animal impact, or a few sacks of clover seed will serve your purpose better than the application of a synthetic fertilizer or abandonment of the salinized area.

Energy Flow: Assessing Sunlight Conversion

Energy flow is hard to observe in a precise way. It's a function of the total area of leaves actively converting sunlight into forage, the length of time this conversion goes on, the efficiency of the conversion, and what happens to the forage after it is grown (see fig. 3-8). Generally speaking, then, high energy flow is characterized by the following:

- An abundance of broad-leafed plants of the "middle" (or mesophytic) variety that can grow rapidly due to a good balance of air and water around their roots
- Close plant spacing
- Rapid growth unimpeded by poor aeration, compaction, or other soil problems
- Plants active through the longest possible growing period (both warm- and cool-season plants in temperate climates)
- A long and robust food chain—herbivores, predators (including humans), decomposers, and so on

Low energy flow is characterized by the following:

- Waxy or narrow-leafed plants that have slow growth rates generally
- Gray, oxidizing grasses
- Wide plant spacing
- Slow plant growth due to capping, compaction, waterlogging, or other soil problems

Soil Capping: Its Development and Effects

As *Holistic Management: A Commonsense Revolution to Restore Our Environment*, third edition, explains in some detail, the capping of bare soil by a hard crust, often cemented by algae and lichen, can inhibit the development of the entire community in brittle environments and block the cycling of minerals into the soil. Though

capping may reduce erosion at the site, it increases runoff that causes erosion elsewhere. Capping may form even on sandy soils, but heavier soils that have lost their crumb structure and lack litter and humus may seal the instant water strikes. Animal impact, either through stock density or herd effect, is the most practical way to break up capping while incorporating organic material into the soil.

Soils covered by litter and living vegetation will not develop capping and usually indicate effective water and mineral cycles and advancing succession.

In monitoring we note four degrees of capping:

   1. Mature capping results from long rest in either form—partial or total. Low-successional lichen, moss, and algal communities often give it a dark tone, as shown in figure 3-9, and it may sound hollow when tapped. Mature capping is seen most often in very brittle environments but is not uncommon in less brittle environments where mosses (and in northern climes also lichens) tend to form the cap rather than algae and lichens. Mature capping in less brittle environments is a healthy sign in the early stages of succession where soil has been exposed, but if succession does not advance beyond moss it may indicate that disturbance is necessary to overcome what is commonly partial rest. In brittle environments, mature capping is almost invariably the last life left in a deteriorating situation.

- *Immature capping* has been broken in the past, as shown in figure 3-10, or scoured by erosion, but has not been present long enough to develop a layer of lichen, algae, or moss. It is still strong enough to inhibit water penetration, aeration, and the establishment of seedlings.

When rest in either form is applied in nonbrittle environments, moss, and in some places lichens, too, will cap the soil in between widely spaced plants. As in brittle areas, animal impact can be used to break this mature capping so that new grass plants can establish.

Energy Flow Illustrated

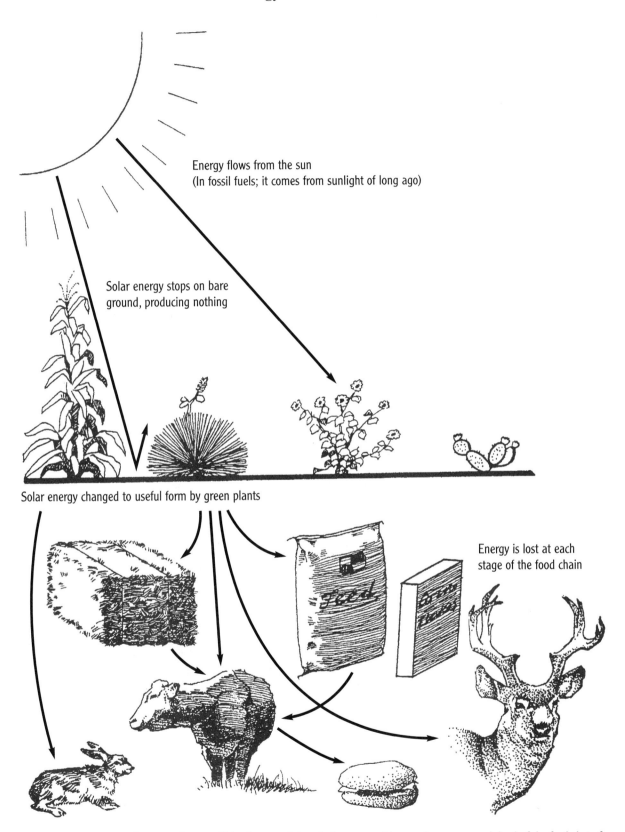

Energy flows from the sun
(In fossil fuels; it comes from sunlight of long ago)

Solar energy stops on bare
ground, producing nothing

Solar energy changed to useful form by green plants

Energy is lost at each
stage of the food chain

Figure 3-8. A mathematical value for energy flow would be the sum of the calories preserved in organic matter at each level of the food chain from an original amount delivered to a unit of land over a unit of time.

Figure 3-9. Mature capping. This surface has been capped long enough for a hard algae-covered crust to develop. Animals had walked across it leaving hardly any trace, but the surface could be broken with a knife, as shown.

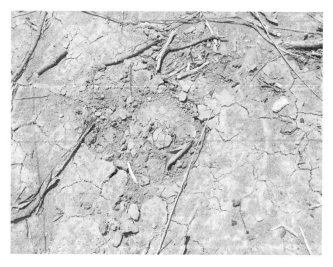

Figure 3-10. Immature capping. This surface has been recently capped by rainfall but has not had time to develop a hard crust (mature capping). A cow has walked across the surface, breaking it easily.

- *Recent capping*, as shown in figure 3-11 on the left, is the result of recent precipitation over broken soil surfaces. It is usually quite thin, depending on soil type, and if left undisturbed becomes immature and eventually mature.

- *Broken capping*, as shown in figure 3-11 on the right, is the result of recent animal impact or other disturbance that opens soil to moisture and aeration, and thus seed germination and establishment.

Figure 3-11. Recent capping (left) and broken capping (right). On both the left and right sides of a fence (not visible), the capping had been broken by a herd of cattle and then recapped following a rainstorm. On the right, the herd of cattle has broken the light capping.

Reading Plant Forms

The form of a plant often tells a history of management and therefore suggests improvements. Notice the following conditions: overgrazed plants, overbrowsed plants, and overrested plants.

Overgrazed Plants

Livestock operators interested in enhancing the productivity of grassland must learn to spot overgrazed *plants*, rather than overgrazed ranges or pastures. Chronic overgrazing of many plants does produce changes that can affect vast areas—the disappearance of certain perennial grasses (often those most productive in a particular season), infestations of unpalatable species, or species adapted to survive overgrazing. But some of the same symptoms also occur when plants (and soils) are overrested, and do not show up until long after corrective action should have been taken.

Especially in situations where low paddock numbers require you to change grazing periods according to the daily growth rate of plants, it is important to continually look for individual plants that have suffered overgrazing, either because of too long a grazing period, or too short a recovery period. Look for these signs (illustrated in fig. 3-12):

- **Distorted growth:** Many plants respond to overgrazing by flattening out below the grazing height of animals. Some species, such as blue grama (*Bouteloua*) and a number of runner grasses, may form into a tight mat, resembling a lawn or golf course that is frequently mown very short. Commonly such grass mats are referred to as *sod bound* with tightly matted but short root systems. Other grass species raise their defenses by hiding new leaf behind spiky stalks formed by old stems.

- **Dead centers:** Bunch grasses subjected to overgrazing, in which they lose energy reserves in stem bases or crowns, will suffer root loss and commonly die back, usually at the center. Overrest, too, can kill plant centers, but in this case the old oxidized leaf usually remains unless subsequently removed by fire or termites.

- **Disappearance:** Some plants very sensitive to overgrazing merely disappear, and remnants are found in sites animals cannot reach.

Overbrowsed Plants

Overbrowsing is closely akin to overgrazing, but it affects perennial *nongrass* plants—shrubs, trees, vines, forbs, and so on. Look for these signs (illustrated in fig. 3-12):

- **Distorted growth or hedging:** Repeatedly bitten branches often develop knobs at the end where new sprouts make a dense cluster. (Gardeners exploit this trait to create hedges by repeated clipping.) Species with a low growth habit may have knobby stems much thicker than normal for the amount of visible foliage. Straight-growing plants such as young pines may split and take the form of bushes (thereby ruining forever the possibility of harvesting timber). Leaves may hide behind spines or old twigs or lie flat against the bark.

- **Browse lines:** Trees lose all foliage below the reach of animals and look like they were trimmed for the benefit of strollers in a park. Some plants may show the knobs and bristles of overbrowsing on their lower branches but long plumes of growth above the reach of animals.

- **Disappearance:** Forbs such as the creeping legumes, often found in grasslands, disappear fairly quickly when overbrowsed. Any species that is being overbrowsed is unlikely to successfully establish young plants.

Overrested Plants

Overrest, followed by overgrazing and overbrowsing, most frequently causes degeneration of brittle-environment grasslands. Successional shifts from grass plants toward woody and herbaceous species (weeds) often result. In brittle environments, dead or dying grass plants tend to produce widening plant spacings and loss of organic matter. You should spot the problem *before* the following changes occur:

- **Old growth that remains standing** into the next growing season or longer—becoming gray or even black in severe cases

- **Plants with dead or weakened centers** that have obviously not been grazed recently

- **Weakened root systems** on plants that have obviously not been grazed (old growth is present)—these dead plants can often be pulled up easily by hand

Identifying Species: It's Not Just a Name Game

How much work should you put into learning the names of plants and animals? Some people memorize names but know almost nothing about the plants and how they function in the community. Like knowing the names of every person in the community but not knowing who is a doctor, lawyer, teacher, or police officer. It is far better to be observant and learn as much as you can about plants and animals and what roles they play in the community.

All organisms fill certain niches within the communities they inhabit. The more you discover about them and how they fit together, the more you will understand the dynamics of your land. To help you distinguish one plant from another in order to monitor them, you will of course need names—but they are just the doorknob to more useful knowledge.

How Overgrazing and Overbrowsing Can Affect Plant Form

Some overgrazed grasses and many runner grasses form dense mats.

Browseline – nothing is growing below the height animals can reach.

A healthy bunch grass plant with time to regrow will have dense lush foliage and a natural shape.

Overbrowsed and hedged perennial shrub of great age.

Prolonged overgrazing creates dead centers and prostrate growth round the edge of some grass plants.

Overrested plants commonly have gray shocks of old growth and dying centers.

Figure 3-12. Overgrazing and overbrowsing can often be detected by distortions in plant growth forms.

Identifying plants from a field guide or key without long practice is a slow and tedious business and misses the point. Get some experts in birds, insects, mammals, and plants to tour your land with you. Keep notes. Ask questions about what various species need in order to thrive, how they reproduce, what preys on them, and so forth.

Though it might take a lifetime to learn a whole biological discipline, you can make a good start on your own limited habitat in a very short time. From then on you can teach yourself what you need to know by your own observations, discussions with people in a learning support group, or a Savory Institute Accredited Hub, and occasional correspondence.

Grazing Patterns and What They Can Tell You

Keep an eye out for any patterns that develop in your paddocks as a result of your animals having been there.

Low Stock Density

Extremely severe grazing was covered in part 2 as a sign of possible overstocking. Aside from the more complex matter of assuring nongrowing-season reserves, you have immediate problems if your animals gnaw grass right down to the quick and pick up litter during the growing season. If this happens in paddocks that also contain untouched, rank forage, the problem is probably low stock density.

If livestock are spread too thin to affect many plants or areas in one grazing, untouched areas may be inedible by the time they return—thus increasing pressure on the remaining space. After several grazing cycles, a large proportion of a paddock may drop entirely out of production. In less brittle areas, the ungrazed patches may advance well along the successional path back to forest.

The signs are easy to spot (and are sometimes called "patch" or "all-or-nothing" grazing):

- Sharply defined ungrazed patches, large or small, of separate species or the same species, as shown in figure 3-13

- Extreme grazing of other areas that in nonbrittle environments often acquire the clipped look of golf course putting greens

Figure 3-13. Low-density grazing. Heavily grazed plants (foreground) border a patch of ungrazed plants. Typical of land showing a low-density grazing pattern, the plants in the background are old and mature and will begin to oxidize, as they are not being grazed, while plants in the foreground are severely grazed. In the foreground, so many plants are overgrazed that soil is exposed and capped, making it more difficult for any new plants, other than weeds, to establish.

In low-rainfall brittle environments, low-density grazing patterns are often inevitable and even desirable. Since the old growth is often not very fibrous, usually cures well, and is an important source of off-season forage, the patches disappear before the next year's growth.

In brittle high-rainfall areas and nonbrittle environments, the overgrowth is typically fibrous, low in nutrients, and must be mowed, burned, or trampled to maintain grassland. If fencing rather than herding is used, the extreme stock density necessary to control this problem in some areas astounds many people new to using animal impact as a tool for creating a desired landscape.

Habits and Routines

Livestock learn routines rapidly—and have a knack for training stockmen to follow them. Sooner or later cows discover that by bawling loud enough at the paddock gate, someone will assume they are starving and move them to the next paddock. In times of drought, especially, this can lead to disaster, because you may have to stress animals slightly to keep the recovery periods long enough. Giving in to them and knocking a day or two off each grazing period can greatly reduce forage and play havoc with your plans and wallet. Faster moves mean overgrazing, which means less production, which raises pressure for faster moves and yet worse overgrazing in a downward spiral.

A Routine Grazing Pattern

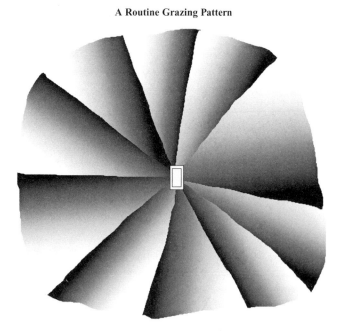

Figure 3-14. Counterclockwise moves have trained the herd to crowd the fence in anticipation of moves.

More subtle routines can also cause problems. If livestock know they will always move into an adjacent paddock, they soon get the habit of crowding that fence line in anticipation. Stock rotated mechanically around a radial grazing unit frequently produce a pattern of severe grazing along one side of each paddock as shown in figure 3-14.

Trailing is another result of habit. Sometimes the placement of fences causes trails to develop, but trails also result from routines that you can break by moving stock through different gates, changing the location of supplements, moving a section of fence, or using temporary electric fence as a diversion. The point is this: Look for signs of destructive routines, and vary them in your grazing planning.

Living Organisms: Community Dynamics Revisited

Monitor living things that help or hurt you and try to understand them in the context of community dynamics. This perspective enables you to use living organisms as a tool—which is nothing more than influencing the biological community to your advantage. This is perhaps the most important and least understood concept in Holistic Management. Changing the environment by direct action (planting, exterminating, stocking, and the like) more often than not fails for trying to produce effects without considering causes.

Holistic Management explains how such headaches as noxious plant invasions, insect outbreaks, and poor soil aeration will not yield unless you correct the distortion in the biological community that produced them. The possibilities are limitless. Informally monitoring small aspects of community dynamics can often lead to big advances in management (see box 3-3).

Box 3-3. A Deeper Look at Living Organisms

There are many examples of creative innovation in land management—from turning pipe leaks into habitat for grasshopper-eating toads to leaving strips of uncut hay for quail cover. For a good expression of the spirit of the idea, here is a passage from Oregon rancher Dayton Hyde's marvelous book *Don Coyote*. After closely observing a series of events on his land, Hyde rejected a common policy of draining marshes and poisoning predators and rodents. Instead he restored wetland and fed coyotes. In his words:

The ancient marshes were coming back: coarse vegetation sprang up almost overnight, creating a whole new habitat for wildlife—no matter that those grasses did not rate well in the county agent's handbook. Cows got fat on them, and that was what paid the bills. All day long, the tea-colored marsh waters absorbed the rays of the mountain sun, and all night long they released heat, warming the climate. Where once even sedges and rushes had turned brown with the night frosts, now they grew green and lush, and frogs croaked all night in the warm, moist darkness. For once we had more feed than cattle to eat it.

In those same marshes, a host of birds—yellowthroats; red-winged, yellow-headed, and Brewer's blackbirds—raise their young. We helped them by restoring the marshes; they helped us by taking care of insect problems. The great grasshopper epidemics that had plagued my uncle and me became history. In the traditional grasshopper nesting areas, clouds of birds worked on nymphs, carting them away to feed their young. Hour after hour, day after day, sandhill cranes stood shredding young grasshoppers with their long bills.

We had once relied on poisons; now as we checked the areas where grasshoppers had once turned the land to dust, we found blue-birds, wrens, flickers, sparrow and red-tailed hawks, coyotes, badgers, foxes, cranes, black-birds, meadowlarks, and even ducks cleaning up the areas. By the time their young were raised, the 'hoppers they'd missed wouldn't have filled a tobacco can.

There is more to maintaining wetlands than backing up water and letting it stand. Marshes tend to choke up with coarse vegetation, which, left ungrazed or unharvested, rots. Periodic drying is essential.

The balance of wildlife species too is important. When we lost our coyotes to poison, the raccoons proliferated until there was hardly a nest on the place they hadn't destroyed. Before the coyotes remultiplied and were again numerous enough to control the raccoons, waterfowl production had been almost nil. Now we had ducks and geese back nesting by the hundreds.

I thought of other species on the ranch. Without flickers, badgers, trout, deer, or chipmunks, the ranch still would have flourished. But if I took away the coyotes, the whole system fell apart. They were as necessary to the well-being of Yamsi Ranch as any tool I owned, including shovel, pick-up truck, mower, hay baler, fencing stretcher, pliers, welding outfit, saddle horse, saddle, rope, medicine, and tractor. In fact, if I were to design a kit for the beginning rancher, a pair of coyotes would have to be included.

There was a difference between my ranch and every other I knew about whose owners complained about coyote damage. Acre for acre, I had three to five coyotes to their one. Yet while they lay awake at night waiting for predators to kill their livestock, I slept like a baby, not hoping, knowing my calves were safe. The secret, of course, was that I kept my coyotes fed all year round.

Ever since the great mouse epidemic, I had relied less and less on poisoned grains to keep ground squirrels in check, and now I used none at all. The epidemic of these little grass eaters I had feared didn't happen. Populations stayed low and fairly stable, and I came to look on the ground squirrels I had on my land as beneficial, since they fed my varied predators.

Monitoring Your Land

THE FOLLOWING PROCEDURES FOR PERIODIC MONITORING of selected sites were developed over a number of years with the input of numerous holistic managers and educators. Although they contain elements from other monitoring techniques, these procedures put special emphasis on soil surface conditions and plant density. If your environment leans toward the brittle end of the scale and much of your land is down to bare ground (between plants), that is where you will get the earliest warning that a shift in succession is beginning, which is far more important than knowing that a shift has already occurred. If your environment leans toward the nonbrittle end, the soil surface will not remain bare, but it can be capped with moss and lichen that prevent succession from advancing to grassland. So, in these less brittle environments soil surface conditions are still important to monitor, but the focus is more often on plant diversity. In every case, monitoring should always look as far down the road as possible for any curves that might require action on your part as a manager.

If you have not ever done it before, taking the responsibility for monitoring your land yourself should symbolize a significant shift in your whole approach to management. Almost all stock growers weigh their animals at least once a year. When you realize that the stock functions only as a broker in the marketing of solar energy, it makes more sense to "weigh" the primary agent in this transaction—your land. And just as you would never consider calling in a stranger to weigh your stock, it might be a good idea not to trust the monitoring of your land to anyone but yourself. That said, there is value in also having a third party monitor your land, which adds another level of reliability and validity to the results.

The Monitoring Procedures

Two monitoring procedures, basic monitoring and comprehensive monitoring, are presented here. *Basic monitoring*, the first, and simplest, procedure involves taking photos of fixed plots and making notes of what you observe. This is all most managers require as a routine monitoring procedure, and it is also handy for monitoring test applications, or "mini-trials," of different tools or techniques in test areas (box 3-4, gives an example). It should not take more than a day of your time each year, about forty-five minutes per transect (depending on your observation skills), a few hours or less for a mini-trial.

Comprehensive monitoring, the second procedure, is, as its name implies, more comprehensive and is ideal for monitoring public lands that require more detailed records, or for those who simply want more detail on which to base their decisions. This procedure takes a few days of work once a year.

A combination of the two procedures may work best for some. The basic monitoring does not produce "quantitative" data. That is, you do not get the satisfaction of learning that your bare ground has decreased by a certain percentage over a period of time, or that your perennial grasses have increased by a certain percentage. The

Box 3-4. Mini-Trial (Early-Warning) Photo Monitoring

If you're placing a transect in an area to test a new action you're contemplating, you may choose to take photos only, with or without the PVC frame, and dispense with filling in the forms. It all depends on what you're trying to learn and how much information you need to convince yourself of the trial's success or failure.

In the photos shown here, all that was done for the trial was to take fixed-point photos of an area before and after receiving ultra-high animal impact for a week to break up some very hard capping. Animal impact was

the tool on trial to see if it could have an effect on land that had been bare and hard-capped for decades. If you want to do something similar, the key is to make sure you have fixed features in the camera viewfinder so there is no doubt that you're looking at the same piece of ground.

Box 3-4b. The same area six months later at the end of the growing season in a drought year—eight inches (200 millimeters) of rain received in a thirty-inch (750 millimeter) rainfall area. A day after the first photo was taken, a herd of 200 cattle and just over 100 goats spent each night for one week bunched in a movable predator-proof enclosure, and then left the area. The resulting spurt of growth, in a drought year, was all the manager needed to see to be convinced that this and other hard-capped areas could be healed with very high animal impact (herd effect). It also ruled out the possibility that some soil factor was inhibiting growth, or that nearby trees were poisoning the soil with toxins released by their roots.

Box 3-4a. This land, managed by Zimbabwe Savory Hub, the Africa Centre for Holistic Management, had been capped hard and bare for several decades and was believed by many to be beyond reclamation. A herd of between 300 and 600 cattle and goats had made no impact in four years. This photo was taken at the end of the dry, nongrowing season.

comprehensive procedure does produce these data. A good compromise is to use the comprehensive monitoring procedure in the first year as a baseline, and then again every five years, or in intervening years if you run into trouble that more detailed monitoring might help rectify. At all other times use the basic procedure.

The data yielded by systematic use of these two monitoring techniques do not serve as a substitute for the day-to-day monitoring of growth rates or the general awareness of events that is required to implement a holistic grazing plan. In some ways, however, the infor-

mation is more important because it generates a cumulative photographic and written record that quickly reveals trends.

The data you gather from your monitoring, even when using the comprehensive procedure, are unlikely to satisfy researchers or academics, who will find them too subjective, statistically invalid, lacking replications or comparisons to a "control," and any number of other things. But their approach to monitoring is from an entirely different perspective. You, remember, are monitoring to bring about the change you intend, not merely to record what happens. The main value of monitoring to you is that at least once a year you are compelled to closely observe the source of your livelihood—your land.

Gathering the Data

Before you get started you need to know where, when, and what you will monitor and why.

About Transects

The word *transect* means crosscut—the theory being that samples taken from selected crosscuts of land, voters, or any other entity too big to test in its entirety will yield a good approximation of the truth. The monitoring procedures described here use different sampling techniques. The basic procedure uses straight-line transects with fixed sampling points, five along each transect. The comprehensive procedure uses a fixed-area transect with random sampling points—up to a hundred—located by throwing a dart backward over your shoulder. The transects, or sampling areas, should be sited in a representative area where you most want to produce change. Three to five transects should be adequate on fairly uniform land. More will be needed if you have many different soil types or varied terrain.

Time of Year to Monitor

Routine monitoring, such as this section describes, should be done at the same biological time each year. The exact day is not critical, since seasons shift from year to year. Bear in mind that monitoring yields the best information during the most active part of the growing season. It's the best time to identify many plants and observe their characteristics. Biological activity is normally highest as well.

In some areas, such as those in Mediterranean climates, heaviest rainfall occurs long before the peak of the growing season; it may include the period when plant cover is typically at its lowest. Dense vegetation during the most active growing season may seriously mask the peak rainy season's soil surface conditions. It may be wise to monitor such areas twice each year to spot the earliest signs of change. Soil cover and movement may be best monitored at a very different time than biological complexity.

In other areas, often the case in the tropics, the best time to monitor soil surface conditions when bare ground between plants is extensive, may be at the end of the dry, nongrowing season. That's when the amount of bare ground is most visible.

Try to have the same person collect the data from your transects each year. If you must use someone new, work with him or her ahead of time to standardize the procedure. New people will inevitably change procedures slightly, which could result in misleading data, but you can minimize the problem.

Monitoring Criteria May Change

The criteria you emphasize most in your monitoring may well change as the land improves. If you manage to produce a dense perennial grass sward on one section of your property, it becomes pointless to monitor the spacing between plants and soil capping there, because there is unlikely to be any. You might turn your attention to increasing energy flow in that section and choose to monitor for an increase in broader-leafed plants as you strive to achieve that change.

Completing the Feedback Loop

Monitoring is useless if you do not complete the feedback loop and take action when you note adverse changes or no change at all. Remember that you are monitoring in order to make happen what you want to happen, to bring about desired changes in line with your holistic context. The final step in each of the monitoring procedures presented here reminds you of that and asks you to record the action you plan to take. There will, of course, be times when circumstances force a delay. Inaction then is a conscious decision and one that will be rectified.

If you think through the tools you have used and how you have applied them, the action you need to take usually becomes obvious. If it does not, reread chapters 15 through 23 of *Holistic Management* (third edition), to refresh your memory of how the tools described affect water and mineral cycles, community dynamics, and energy flow. If you are still uncertain but one or more possible actions come to mind, test them out on a small piece of land, as described earlier in box 3-4. Alternatively, take another approach altogether and look for evidence elsewhere on your property where changes are moving in the direction you want them to go. In other words, look for positive deviations, as described in box 3-5.

Basic Monitoring

Although simple, this monitoring procedure is effective, in that with minimum work it allows you to observe changes closely and also provides a good record of those changes

Equipment Needed

You will need the following equipment:

- Camera (preferably one that takes high-resolution photos)
- 100-foot (30-meter) metal tape
- One square yard (or meter) frame of lightweight PVC
- Pad of paper
- Clipboard and pen or pencil
- Monitoring data forms (five per transect)

Box 3-5. Looking for Positive Deviations

Some years ago, Allan Savory wrote an article about ranchers in low-rainfall brittle environments who had found their land to be stagnating—no change was occurring where change was planned. He said the ranchers hadn't taken action to overcome the stagnation mainly because they didn't know what to do.

After diagnosing the cause of the stagnation, Allan went on to stress the importance of monitoring and changing course at the earliest sign of stagnation and offered some pointers to help ensure this actually happened. But being aware that there is a problem is different from knowing what to do about it.

Allan apparently figured out what to do (or confirmed his hunch) in each case by looking on the land for exceptions to the general trend, or positive deviations. On each of the ranches he visited he asked if there was any place where things had not stagnated. Some initially said no; others immediately said yes. In fact, when he went out and looked, they all had pockets of improvement somewhere, and in each case these areas were located either in the smallest paddock on the ranch, or where livestock had been periodically crowded.

In each case the land had continued to improve where animal impact was repeatedly highest. So Allan recommended that the farmers each take steps to increase animal impact on the rest of their land.

To choose his management solution, then, Allan relied on the positive deviations he found by looking around the land, rather than on data from year-to-year monitoring. Indeed, the information he needed could easily have been missed if he had relied solely on the transect data, since the transects are only small samples of the land.

A way to remedy the situation highlighted in the article, would be to make it standard practice to carry out two different kinds of monitoring:

1. *Monitoring for overall progress* through fixed-point monitoring repeated year after year in the same places. It addresses the question, Which way am I moving?
2. *Monitoring for positive deviations*, or "free range monitoring," sends you all over the land (even beyond your property boundaries) looking for exceptional improvements wherever you can find them. When fixed-point monitoring shows you're moving in the wrong direction (or not moving at all) the positive deviations observed in your free-range monitoring may give you the clues you need to answer the question, How can I move in the direction I want to go?

— *Rush Wayne, Oregon, USA*

When you are establishing the transects, you will also need these items:

- Substantial metal posts, or heavy-duty plastic stakes, to be driven well into the ground at either end of a line as permanent markers (two per transect).

- Five short lengths of rebar rod (about a foot, or 30 cm long), or heavy-duty plastic survey "whiskers" with spikes (if rebar is used, you may want to add a magnetic stud finder to the preceding list).

- Heavy hammer adequate for pounding rebar rods fully into the ground if rebar is used.

You will be establishing straight-line transects, along which you will have five monitoring plots contained within a square frame. You are going to be standing over the square frame taking photos straight down. This is not always easy while trying to keep your shadow out of the photo, particularly at midday. Orienting your transect line east–west may help, and certainly for the longer-distance photos you will also take, but to avoid the shadow problem altogether, consider using a modified camera stand. If you are handy in the workshop, you can make a stand to hold the camera at the correct height directly over the plot (see fig. 3-15). This can be made out of light angle iron or galvanized pipe, threaded at one end or with a magnet so the camera can be attached.

Establishing Transects

You will need to select your monitoring sites carefully. Pick areas to sample that are either typical of the whole area or where you particularly want to produce a lot of change. The more uniform the land, the fewer sites you will need. On uniform units, a minimum of three to five transects give good information.

At each site establish a transect line by hammering the first post well into the ground. Steel posts are best for stony ground. If your soil is sandy, or of high shrink–swell clay, longer steel fence posts that can be driven deeper into the ground might serve better. The heavy-duty plastic stakes used by surveyors work well in other areas. Make

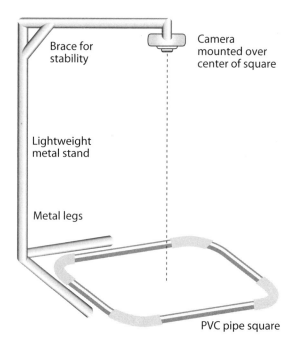

Figure 3-15. A homemade camera stand such as this one, made of metal, makes it easy to take waist-level photos with the camera directly over the center of the plot, and helps eliminate shadow.

sure posts extend high enough above ground to be clearly visible to anyone traveling on a four-wheeler or motorbike. To discourage animals from rubbing against posts and dislodging them, try piling stones around them.

Once the first post is in place, string out the 100-foot (30-meter) tape to the east or west, and hammer in the second post. Leave the tape stretched between the posts.

Next, go along the tape and at five equally spaced measures—10, 30, 50, 70, and 90 feet (5, 10, 15, 20, and 25 meters), for example—hammer in a short rebar rod flush with the soil surface. The tape will help you relocate the rod each year, but you may also need a stud finder. Alternatively, you can mark each spot with survey whiskers—brightly colored plastic frills that protrude about 5 inches (12 centimeters) above the ground—after you have secured them with an 8-inch (20-centimeter) spike. Their color will fade over time, but they generally provide an effective marker for several years.

Fill in the required information (except photo numbers) at the top of the Basic Ecological Monitoring Data form (see fig. 3-16). You will need five forms per transect, one for each plot. Assign a number to the transect (for example, 1) and to each plot (for example, 1-1, 1-2,

ECOLOGICAL MONITORING DATA - BASIC

(5 Needed per Transect)

Property Harris Ranch   **Transect/Plot #** 1-2   **Photo #s** 1-2

Date 6-21-18   **Examiner(s)** John, Amy

1. **Soil Surface**
    Describe the nature of the bulk of the soil surface between plants. (Is it bare, capped, broken, covered with litter, covered with algae and lichen, hard, soft, porous, etc? Are there signs of soil movement/erosion, such as pedestaling, siltation in low points, etc.?)

    > Most of the surface has recently broken capping & fair amount of litter. Still some patches (+/-5%) of mature capping.

2. **Animal Sign**
    What signs of animal life are present (small or large animals, birds, insects, reptiles)?

    > Lots of insect sign, mostly ants. Rabbit droppings.

3. **Litter**
    If there is litter present, describe its quality/condition (fresh, old. or breaking down so it is hard to distinguish where litter ends and soil begins.)

    > Litter is all fresh—all from this year.

4. **Perennial Grass Condition**
    if perennial grasses are present, describe their condition. (Are they healthy, mature, young, seedlings, dead/dying, overrested, overgrazed?)

    > Most plants now healthy (5), one is dying from overrest (gray).

5. **Grass Species**
    List grass species in the plot if you know their names.

    > Little Bluestem.

6. **Other Plants**
    List or comment on other non-grass plant species present (legumes, forbs, etc.)

    > One new legume of unknown species. A few weeds (species unknown).

7. **Points of Interest**
    Note any other points of interest, including things that might not show well in the photo.

    > The overrested plant is the same species as the healthy ones—Little Bluestem!

Figure 3-16. Complete this form while out on the land.

etc.). Note the date and the name of the person recording the information. Use the back of each sheet to record any other information you feel you need to record.

If forms are shuffled out of order or evaluated by someone else, lack of proper identification will render them meaningless. Complete information will ensure ready retrieval from whatever filing system you use to store photos and forms.

Taking Photos

Start by taking two photos, one from each end of the transect line, to show the general view in each direction. Each photo should include one-third sky and two-thirds foreground. Before you store them, label the photos using the transect number and the direction in which the photo was taken. For example, if the line for transect 1 runs east to west, the photo taken from the west end looking east could be identified as 1-E (eastern view) and vice versa. Note the photo information in the blank area at the top right-hand corner of the first monitoring data form you fill in. If there are no fixed features in either of the two photos and you risk being unable to identify them later, write the identifying detail (1-E, etc.) and date on a large piece of paper that can be tent-folded so it stands up. Two dry-erase boards hinged together with straps and set upright like a tent also work well. Place the paper or board within the camera's field of vision and close enough that the writing is visible.

Next, at each marker along the transect, lay down the PVC frame so that one corner is sitting over the rebar peg or plastic marker and one side is flush with the stretched tape. Write the plot number and the date boldly on a piece of paper or dry-erase board, large enough that the letters will show clearly in the photo. Place this piece of paper or board in one corner of the frame, as shown in figure 3-17. (You will need to position the frame and the piece of paper or board in exactly the same place each time you retake the photos).

Take a photo of each of the five plots along the transect with the camera directly above the center of the plot and at such a height that the entire PVC frame is within the view of the camera. This is where a takedown stand that holds the camera will be useful. If it has only one upright, as shown in figure 3-15, it will be relatively easy to keep any shadow out of the picture.

Figure 3-17. Photo monitoring frame. Frames constructed with PVC pipe can easily be taken apart for storage. The paper or dry-erase board listing the identifying information should be placed in the same spot year after year.

Recording Your Observations

After taking the photo at each plot, note the photo number at the top of the basic monitoring data form. Then record your observations as follows:

- **Soil surface:** Describe the nature of the bulk of the soil surface between plants. Is it bare, capped, broken, covered with litter, hard, soft, porous, covered with algae, lichen, or moss? Are there signs of soil movement (erosion) such as pedestaling, siltation in low points?

- **Animal sign:** What signs of animal life are present? Large animals will leave hoof or footprints, scratch marks, and dung. Smaller animals might leave droppings, minute trails, or signs of burrowing. Birds and reptiles will leave some of those signs, as well as nests, feathers, and burrows. Insects will too, and may also be sighted. Earthworms may leave castings behind.

- **Litter:** If there is litter present, describe its quality and condition. Is it fresh, old, or breaking down so that it is hard to distinguish where it ends and soil begins?

- **Perennial grass condition:** If perennial grasses are present, describe their condition. Are they healthy, mature, young, seedlings, dead or dying, overrested, overgrazed?

- **Grass species:** List grass species in the plot if you know their names. List both annuals and perennials. If you do not know its name but think a grass is an important indicator, take home a sample and get help from an extension adviser or other specialist who can identify it. Or you can simply note whether the grass is a cool- or warm-season variety if you live in an area where a mix of the two would be a notable factor.

- **Other plants:** List or comment on other non-grass plant species present. Are there any forbs, such as legumes, that help mineralize the soil, noxious plants that could be poisonous to stock, or other nongrass plants that provide feed at critical times? Are they annuals, biannuals, or perennials?

- **Points of interest:** Note any other points of interest, including things that might not show well in the photo. You may see something that interests you but is not covered in any of the above points. Or perhaps you see a trend developing that you want to keep your eye on.

Analyzing Your Observations

Remember that the primary purpose of this monitoring is to make happen what you want to happen. Based on what you observed at each of the photo plots, you can begin drawing conclusions on where you stand relative to your holistic context and what action you need to take.

Set Up the Monitoring Analysis Form

Fill in one Basic Ecological Monitoring Analysis form (see fig. 3-18) for each transect. Start by recording the details at the top of the form. If your files ever become separated, there should be no doubt about which data forms are covered by which analysis form.

Record Your Analysis

Review each of the monitoring data forms you filled in for the transect and summarize your findings relative to each of the questions asked on the monitoring analysis form. In the first year, the photos and your analysis will serve as baseline information. In subsequent years, the information you record on the analysis form will be the basis for a great many decisions. Record your answers to the following questions, using the back of the form as needed:

1. **Future landscape:** What are you trying to achieve in the area surrounding this transect? The landscape described in your holistic context should have been expressed in terms of the four ecosystem processes, but fairly generally. Because your transects are likely to be sited on different soil types or within different environments—for example, clay flats rather than riparian areas—a description more specific to the site provides more focus to your observations. In addition, indicate whether you are attempting to create specific landscape features, such as brushy areas. Your answer to this question will likely remain the same year after year.

2. **Progress check:** What progress have you made this year compared to last? Review each of the monitoring data forms to get a sense of where you are now. Note specific positive or adverse changes, or no change at all, in terms of community dynamics, water and mineral cycles, and energy flow. For example, if the bulk of the soil surface on each plot was bare and covered with mature capping, what does that tell you about the water cycle?

3. **Influencing factors:** What natural or management factors might have influenced what you are seeing on the ground? Think in terms of natural forces, a fire or flood that swept through the transect area during the year; weather factors, such as a heavy downpour or a hailstorm that occurred a few days before you made your observations, or a complete rainfall failure. If your stock have been in the paddock covered by the transect very recently, or not for months, that would be worth noting. If you created herd effect with an attractant, the land will be different than it would have been otherwise, and you should note that the herd effect occurred.

ECOLOGICAL MONITORING ANALYSIS - BASIC

(Use 1 per Transect)

Property Harris Ranch   **Transect/Plot #** 1-1 to 1-5   **Photo #s** 1-1 to 1-5

Date 6-21-18   **Examiner(s)** John, Amy

1.   **What are we trying to achieve in the area surrounding this transect?**

> Community Dynamics   Healthy grassland with legumes & other forbs in open country.
>
> Water Cycle   Very effective to restore groundwater.
>
> Mineral Cycle   Effective—100% biological decay.
>
> Energy flow   High

2.   **What progress have we made this year, compared to last year?**

> Community Dynamics   Improvement! Lots of new plants, grass & forbs
>
> Water Cycle   Great improvement—less capping, more liter.
>
> Mineral Cycle   Improved. Far less oxidizing grass.
>
> Energy flow   Improved greatly—healthier plants, more of them green.

3.   **What natural or management factors might have influenced what we are seeing on the ground?**

> Rain average, and good distribution.

4.   *If adverse changes have occurred or no change, where change was planned:* **What is the underlying cause (what tools have been applied, and how have we applied them?)**

> All tools working to plan. If anything, stocking rate still low since there's still a lot of old grass.

5.   **What are we going to change in this next year to keep our land moving toward the future landscape described in our holistic context?**

> Increase stocking rate slightly, as product conversion is still the financial weak link in the cattle operation.

Figure 3-18. You can complete this form when you return home.

4. **Change or no change:** If adverse changes or no changes have occurred where change was planned, what is the underlying cause? What tools have you applied, and how have you applied them? Positive changes that show you are moving toward your goals are important, but more important, because they require immediate action, are adverse changes, or no change at all where you had planned for change to occur. Carefully consider the tools you have used and note how they could have affected the four ecosystem processes. (See the discussion on analyzing the tools in the "Comprehensive Monitoring" section, which follows.) If the soil surface was bare and capped and you increased your stocking rate and reduced the size of your paddocks, and no change occurred, perhaps the tools you used, such as grazing and animal impact, were trumped by rest (in the form of partial rest).

5. **Proposed actions:** What are you going to change in the next year to keep your land moving toward the future landscape described in your holistic context? What you propose to change over the next year as a result of no change or adverse change is critical to making progress. In most cases, taking action will require the use of a tool other than the one that led to the adverse change or no change, or a modification in how you applied the tool. You have identified the conditions and the causes and therefore should now be able to identify the tool required to bring about a remedy. Determine which management guidelines apply and then decide how you will use that tool. When you have completed this step, you will find that you have outlined the actions you should take. Check to make sure those actions are aligned with your holistic context and modify your plans accordingly.

To continue with the example in question 4, you would need to determine how to overcome the partial rest you inadvertently applied if that is what you think was the problem. Animal impact is what you would look to, but not as you applied it last time. Now you would plan to increase it significantly. The management guidelines that apply are stock density and herd effect. *Stock density* refers to the density of animals on a given unit of land at a given time. Subdividing an area with fencing enables you to increase stock density, but rarely is it sufficient to overcome the detrimental effects of rest. *Herd effect* refers to the effect a large herd of bunched animals produces as a result of a behavior change that has a dramatic impact on overrested soils and plants. (Stock density does not change behavior until applied at an extreme level involving a great deal of fencing that crowds animals so closely in a confined space their behavior changes.) To increase herd effect, you could amalgamate herds, use an attractant that would enable you to increase animal impact over small areas, or use herding, which would enable you to increase animal impact over large areas. In some cases, you could use strip grazing with temporary fence to achieve the needed ultra-high density that induces behavior change. (See box 4-4, "A Land Plan for Ultra-High-Density Grazing," in part 4.) The context checks will help you decide which option to pursue. But in the end, you will assume you are wrong, and check to see the next time you monitor.

Once you have completed each of your summary forms, make sure you file them together with their data forms and photos in a way that prevents them from being separated.

Comprehensive Monitoring

This procedure takes a few days of hard work once a year. It requires you to look at the land in more detail, which will enhance your observation skills, and it will yield far more data than the basic procedure. We encourage you to use this procedure if you are managing public lands or in any other situation where "quantitative" data are required by a government agency, absentee owner, or others.

Siting Transects

You will need to site your transects carefully. Pick areas to sample that are typical of the whole area or areas where you particularly want to produce a lot of change. The more uniform the land, the fewer the transects necessary. On uniform properties, a minimum of three to five transects give good information. On properties with several or many different soil types, or varying terrain—mountains, valleys, and riparian areas—you will require more. However, you must balance the time you can invest against the precision of information you need.

You will be gathering data from random points within each transect area. Point sampling has three requirements: (1) an adequate number of points, (2) randomness in choosing points, and (3) points that really are points. (Measuring from a broad mark such as a footprint, you could fudge.)

To ensure that the sample is random, choose the sample points by throwing a dart backward over your shoulder, being careful not to aim it. The dart technique gives you a "dimensionless" point, but it is difficult to apply where plant spacing is tight, a common situation in less brittle environments. You will sample up to 100 points at each site—fewer than 50 may not give reliable results. Sampling 100 points exactly makes it possible to read many of the findings directly from the data as percentages.

The Monitoring Forms

The three-page Comprehensive Ecological Monitoring Data form, whose first page is shown in figure 3-19, has three major divisions. In the first division, you record what the dart point hit (covered or bare ground). In the second, you note what you found within a six-inch (fifteen-centimeter) circle around the dart point. In fact, you can use any size circle as long as you always use the same size. However, bear in mind that the smaller the circle, the more accurate your judgments of soil surface cover will be. In the third section, you record the distance from the sample point to the nearest perennial plant and provide

additional information on that plant. Each division has lines numbered from 1 to 100. If you think you can get away with only fifty points, stop there.

Feel free to modify this form to better fit your own situation. You may choose to skip certain columns or create others that will give you more meaningful data. In less brittle environments, for instance, some managers measure the distance to the nearest legume, rather than the nearest perennial grass, because they already have a dense perennial grass sward and seek to increase legumes to enhance biodiversity and the mineral cycle.

The Comprehensive Ecological Monitoring Summary form, shown in figure 3-20, includes space for recording the totals and averages of all the information included on the monitoring data sheets.

The Comprehensive Ecological Monitoring Analysis form, shown in figure 3-21, provides space to record your analysis of the results.

Blank forms are included in appendix 7.

Equipment Needed

You will need to have the following items on hand when you go out to collect your data:

- Camera (preferably one that takes high-resolution photos)
- Monitoring data sheets
- Clipboard
- Pencils
- Measuring tape or ruler adequate to measure distance to nearest perennial plant
- Fishing weight on a length of cord (if in brushy country)
- Steel posts or suitable permanent markers, three per transect; optional short marker post with distinctive top, one per transect (these are required only the first time when marking the boundaries)
- Bright-colored darts (the heavier, the better) with two- to three-inch (fifty- to seventy-five-millimeter) tips (see fig. 3-22)
- This handbook

ECOLOGICAL MONITORING DATA - COMPREHENSIVE

🖋 Savory

Property: Dillon Ranch Transect: 4 Photo #s: 4-1, 4-2 Date: 6-26-18

Examiner(s): Joe, Julie, Steve

Other Comments:

The form records, for each throw, data under the following groupings:

- **AT DART ENTRY POINT — What Dart Point Hit (Must Check One):** Bare Soil, Litter 1, Litter 2, Rock, Plant Base, Canopy Above Point (Yes)
- **6-In (15-cm) CIRCLE AROUND POINT — Soil Surface (Must Check One):** Mature, Immature, Recent, Broken, Covered
- **Evidence of / Check if 'Yes':** Animal Sign: I,W,B,S,L; Annuals Present; Soil Movement; Grass: (C),(W), or (Y)
- **DESCRIBE NEAREST PERENNIAL — What It Is (Must Check One):** Rush or Sedge, Forb, Shrub, Tree, Distance to it (in/cm)
- **Its Habitat (Check One):** Dry, Middle, Wet, Seedling
- **Its Age (Must Check One):** Young, Mature, Decadent (Dying), Resprout
- **Its Form (Must Check One):** Normal, Overrested, Overgrazed, Overbrowsed, Dead
- **Its Species (If Known)**

Throw Number	Distance to it (in/cm)	Habitat	Its Species (If Known)
1	25	Dry	LEMON PECTIN
2	125	Dry	UNKNOWN
3	60	Dry	" "
4	15	Dry	" "
5	5	Dry	" "
6	65	Dry	PANICUM
7	50	Dry	RED GAMMA
8	50	Dry	BUFFALO GRASS
9	40	Dry	UNKNOWN
10	35	Dry	" "
11	90	Dry	RED GRAMMA
12	60	Dry	UNKNOWN
13	50	Dry	" "
14	10	Dry	CURLY MESQUITE
15	50	Dry	UNKNOWN
16	75	Dry	" "
17	5	Dry	" "
18	20	Dry	PANICUM
19	60	Dry	UNKNOWN
20	15	Dry	RED GRAMMA
21	80	Dry	UNKNOWN
22	25	Dry	RED GRAMMA
23	40	Dry	UNKNOWN
24	40	Dry	RED GRAMMA
25	40	Dry	S. SIDA
26	10	Dry	HAIRY TRIDENS
27	60	Dry	3 AWN
28	20	Dry	S. SIDA
29	30	Dry	K. R. BLUESTEM
30	30	Dry	UNKNOWN
31	25	Dry	" "
32	175	Dry	" "
33	20	Dry	

SUBTOTAL: Bare Soil 10 | Litter 1 13 | Litter 2 0 | Rock 10 | Plant Base 0 | Canopy Above Point 3 | Soil Surface Mature 20 | Covered 13 | Animal Sign 29 | Annuals Present 33 | Soil Movement 29 | Grass 0 | Rush or Sedge 0 | Forb 9 | Distance 343 | Form Normal 13 | Overrested 18 | Overgrazed 0 | Overbrowsed 0 | Dead 2

I = Insect; W = Worm; B = Bird; S = Small animal; L = Large animal; **C = Cool season; W = Warm season; Y = Year-round green

Figure 3-19. Completed Comprehensive Ecological Monitoring Data form. The second and third sheets of the form (not shown) record data for points 34 through 66 and 67 through 100, respectively.

ECOLOGICAL MONITORING SUMMARY - COMPREHENSIVE

Property Dillon Ranch   **Transect/Plot #** 4   **Photo #s** 4-1, 4-2   **Date** 6-26-18

Examiner(s) Joe, Julie, Steve

Type	Plant Name or Species	No - %
GRASSES	PANICUM	3
	SIDE OATS	9
	RED GRAMMA	13
	BUFFALO GRASS	10
	3 AWN	4
	K R BLUESTEM	2
	CURLY MESQUITE	3
	Total Grasses : No. - %	69
RUSH		
	Total Rushes & Sedges : No. - %	0
FORBS	LEMON PECTIN	4
	SPREADING SIDA	5
	UNKNOWN LEGUME	4
	Total Forbs : No. - %	13
SHRUBS		
	Total Shrubs : No. - %	
TREES		
	Total Trees : No. - %	

Soil Surface		%
COVER & CAPPING	Bare Soil	10
	Litter 1	29
	Litter 2	9
	Rock	21
	Plant Base	2
	Canopy	8
		-
	Mature	52
	Immature	1
	Recent	-
	Broken	-
	Covered	45
EVIDENCE	Insects	54
	Worms	-
	Birds	9
	Small Animals	24
	Large Animals	-
	Annuals	91
	Soil Movement	94

Nearest Perennial		%
CHARACTERISTICS	Cool Season Grass	-
	Warm Season Grass	69
	Year-Round Green Grass	-
	Rush or Sedge	-
	Forb	4
	Shrub	6
	Tree	-
	Average Distance (Inches /cm)	45
	Dry	93
	Middle	7
	Wet	-
AGE & FORM	Seedling	3
	Young	32
	Mature	68
	Decadent (Dying)	3
	Resprout	9
	Normal	39
	Overrested	57
	Overgrazed	-
	Overbrowsed	-
	Dead	5

REMARKS

Figure 3-20. All three data sheets (100 points) have been summarized here in the Comprehensive Ecological Monitoring Summary.

ECOLOGICAL MONITORING ANALYSIS - COMPREHENSIVE

(Use 1 per Transect)

Property _Dillon Ranch_   **Transect/Plot #** _4_   **Photo #s** _4-1, 4-2_

Date _6-26-18_   **Examiner(s)** _Joe, Julie, Steve_

1. What are we trying to achieve in the area surrounding this transect?

Community Dynamics Move to healthy grassland with scattered trees.

Water Cycle Improve a lot.

Mineral Cycle Improve a lot.

Energy flow Push much higher.

2. What progress have we made this year, compared to last year?

Community Dynamics Very slow, almost no change but have removed many dead plants—a lot are still overresting.

Water Cycle Slight improvement.

Mineral Cycle Slight improvement.

Energy flow Slight improvement—more healthy grass plants than before and a few new (young) plants have come in.

3. What natural or management factors might have influenced what we are seeing on the ground?

Got 3 inches of rain in February, but not a drop since.

4. *If adverse changes have occurred or no change, where change was planned:* What is the underlying cause (what tools have been applied, and how have we applied them?)

Slow change is probably because animal impact still too low (partial rest), due to large paddocks and too few animals.

5. What are we going to change in this next year to keep our land moving toward the future landscape described in our holistic context?

Keep on with current plan to gradually increase fencing (and animal impact) and build up numbers (to get moreoverrested plants grazed) as finances allow.

Figure 3-21. Your monitoring is incomplete if you don't take the time to analyze the results, as shown here on the Comprehensive Ecological Monitoring Analysis form.

Making a Sampling Dart

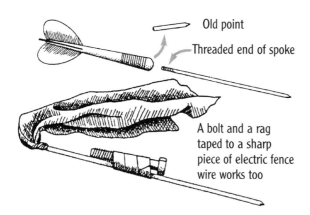

Old point

Threaded end of spoke

A bolt and a rag taped to a sharp piece of electric fence wire works too

Figure 3-22. A toy dart makes a classy tester that penetrates tall grass. Yank out the point. Use a drill to enlarge the hole slightly so you can screw in a bicycle spoke. The spoke should be cut to eight inches (twenty centimeters) and sharpened.

Marking the Transect Area

Locate permanent starting points for each transect. Each transect will consist of up to 100 random points, but the starting point must remain the same from year to year. The transect boundaries are defined with the help of a camera. Standing at the permanent starting point, face toward a fixed feature, such as a mountain, hill, or tree, and center that feature in the middle of the camera's viewfinder. The left and right extremities of the camera's field of vision will define the remaining boundaries, which form a triangle, as shown in figure 3-23. If you can lay out your transect so that the fixed feature is to the north or south (facing north in the Northern Hemisphere and south if you live in the Southern Hemisphere), you reduce your chances of casting a shadow over the photos you will be taking from this same spot each year.

Once you have located the starting points and right and left boundaries at each site, mark them with a steel post, rock pile, or other permanent fixture. If you are managing an extensive area of land, record the location of the starting points on a map and write down the directions for getting there if at all complicated. Mark the routes on the map, if necessary. A GPS can also be used to record starting points, but back up the information on paper.

General Layout of a Transect

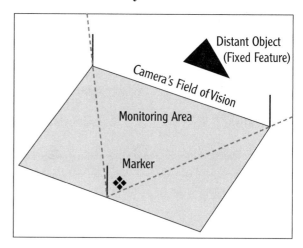

Distant Object (Fixed Feature)

Camera's Field of Vision

Monitoring Area

Marker

Figure 3-23. The transect area, within which you will throw the dart, lies within the triangle formed by the three marker posts. The starting post (bottom) should be located so that the fixed-feature (distant object) lies at the center of the camera's field of vision. If the fixed feature is located north or south of the starting post it may help you avoid casting shadows in the photos you will take from this spot. The "Marker" is used when taking close-up photos of the soil surface.

Recording Transect Information

For each site, do the following:

1. Set up the Comprehensive Ecological Monitoring Data form (three pages). Fill in the identification information at the top of each page (except photo numbers). Labeling each page is important in case pages get shuffled among other sheets.

2. Take photos. First, write the transect number and date on a large piece of paper that can be tent-folded so it stands up. Two dry-erase boards hinged together with straps and set upright like a tent also work well. Place the paper or board within the camera's field of vision and close enough that the writing is visible.

Stand at or against the starting point marker and face the fixed feature you used in defining the transect boundaries. Take a photo directly facing the fixed feature. The feature should be centered in the viewfinder, and the photograph should include one-third sky and two-thirds foreground. Being consistent will help you more easily compare photos of the same site year after year. Try to use an identical camera each year, or at least be consistent with the lens setting (if using a zoom lens) from year to year, so as not to distort the picture of what is happening to your land over time. Taking care when preparing for this first photograph will pay off in the long run, as the fixed point will remain constant in the photos.

Once you have taken the main-view photograph, you need to get a more detailed view of the soil surface in the foreground. Make sure you place the paper or board with the identifying information in the camera's field of vision. There are two options for getting this view:

a. Remain standing where you are and point the camera downward at about a forty-five-degree angle.

b. For a closer, more detailed view of the soil surface, walk a short distance away from the marker toward the fixed feature. (Note: You should settle on a distance for this, which you will use for all your monitoring sites—somewhere between five and ten yards (or meters)—so that you are away from possible influence of livestock concentrating at the main marker. At that point, drive in a short marker that has a distinctive top, such as the plastic whiskers that surveyors use, or a metal washer or painted piece of metal welded to it.

Take a photo, from waist-height, with the camera pointed straight down to record what is happening at ground level. Use the head of the marker for this spot as a reference by locating it in the center of the viewfinder.

Give each photo a number before you store it, and record those numbers at the top of the monitoring data sheets. As soon as you print your photos, make sure they are numbered too.

3. Throw the dart. Stand at the permanent starting point and toss the dart backward over your shoulder anywhere within the transect area, making a conscious effort not to throw the dart where you hope it will land. Painting the dart fluorescent orange or tying on a piece of orange flagging may help you to find it in dense cover. Short tosses are easiest to find.

When you finish documenting one sample point, as described in the next section, move on to the next, taking care to spread your sample points so they cover the entire monitoring site by the time you have thrown the dart either 50 or 100 times. The best reason for stopping at 50 throws is if the community you are sampling is fairly uniform.

Some managers find it useful to walk around the transect area before throwing the dart, recording the names of familiar plant species and other outstanding features. Because of the random nature of the dart-throwing technique, some species or soil surface features may not show up in your data.

Recording Your Observations
Use the monitoring data sheets to record your observations at each point, completing a row per throw, from left to right across the page, as shown in figure 3-19.

At the Dart Entry Point
If the dart did not stick, sight straight down over the point to the spot below and record what lies there. Put a check mark in the appropriate column: "Bare Soil" and "Rock" speak for themselves. "Litter 1" refers to a new, undecayed layer of litter (leaves, sticks, dung)—sometimes the "layer" is only bits of twigs or dead leaves on bare ground. "Litter 2" indicates deeper litter that is decaying and being incorporated into the soil. In less brittle environments, the litter often merges into the soil without a distinct boundary. "Plant Base" refers to the area actually covered by the root crowns or stems of plants.

Locating a Point with a Plumb Bob

Figure 3-24. When the dart gets caught in aboveground vegetation, use a plumb bob to locate the point.

In most bunchgrass-dominated communities, plant base, or basal, hits rarely exceed five to fifteen percent. In areas with sod-forming grasses, individual basal areas are hard to distinguish; you must devise a standard. One way is to record a basal hit only if the point falls into a living clump. Otherwise, record litter or bare ground and measure the distance to the nearest live shoot of the sod-forming grass.

If the dart lodges in vegetation above ground, use the fishing sinker and cord as a plumb bob (see fig. 3-24) to find a point on the soil surface directly below and treat that as a hit.

If the dart point has landed immediately under a canopy of any sort, such as a leaf or branch, that would have slowed a raindrop before it hit the surface, make a check in the "Canopy Above Point" column.

In a Circle around the Entry Point
Document the following elements in a six-inch- or fifteen-centimeter-diameter circle centered on the entry point of the dart:

1. **Record soil surface conditions.** Make a check in the column that best describes the soil surface in the bulk of the circle:

   - *Mature capping:* Algae, lichen, or moss covers the surface.
   - *Immature capping:* The surface doesn't have lichen, algae, or moss, but is still sealed and impenetrable; it may have been broken in the

past, or scoured by erosion, but is still strong enough to inhibit change.

   - *Recent capping:* The surface has recently been sealed over by rain.
   - *Broken capping:* The bare surface has been broken recently by animal impact or some other form of disturbance but has not yet formed a distinct cap.
   - *Covered:* Most of the raindrops falling within the circle would land on living or dead plant material, rather than bare or capped ground.

   Note: If the surface within a significant number of circles is likely to be sod bound with low successional grass species, such as blue grama, couch, or Bermuda (*Cynodon dactylon*), record the surfaces as "Covered" but make a note in the "Other Comments" column so you can make a better analysis later.

2. **Record evidence of large or small animal activity.** If you see signs within the small circle of large or small animal activity—tracks, droppings, burrows, mounds, worm castings, or actual sightings—make a note in the "Animal Sign" column: Use *I* for insect, *W* for worm, *B* for bird, *L* or *S* for large or small animal. (If you can identify the species, note it in the "Other Comments" column.)

3. **Note if annuals are present.** Make a check mark in the "Annuals Present" column if an annual grass or forb is growing within the circle. If you know the name of the species, record it in the "Other Comments" column. If you are monitoring what has become largely annual grassland, you could ignore this, since annuals will be everywhere, and you are probably more concerned with the nearest perennial plant.

4. **Record any evidence of erosion.** Put a check in the "Soil Movement" column if you see any signs within the circle of erosion by wind or water—plants or rocks on pedestals, water-flow patterns, litter banks, siltation in low points, or splash patterns.

The Nearest Perennial Plant

Locate the perennial plant nearest to the dart tip. Sometimes several plants of different species might be intertwined. If they have seed heads, you can distinguish them more easily. Note: Where you are specifically trying to create perennial grassland, you might want to locate the nearest perennial grass rather than a woody plant. In cases where perennial plants are scarce, you may find that each time you throw the dart the nearest perennial to the dart tip is the same plant. If so, make a note of it.

1. **Note the type of plant.** Make an entry in the appropriate column for any of the following observations:

   * *Grass:* Indicate with a *C* or a *W* whether it is a warm- or cool-season grass, or a *Y* if it is green year-round.

   * *Rush or sedge:* Sedges resemble grasses but often have solid, angular (rather than round) stems; rushes have hollow, pithy stems.

   * *Forb:* Any flowering plant, other than grass, that does not develop woody stems, for example, a legume, such as clover.

   * *Shrub or tree:* A shrub could be scrubby tree growth. Fix your own definition for what is a tree and what is a shrub, according to the expected growth form of the species in your area.

2. **Measure distance from the dart point.** Using a tape measure, note the distance from the dart point to the base of the perennial plant and record that in the "Distance to It" column. If the dart hit the base of the perennial plant you have identified, record the distance as zero. Note: In sod-bound communities, or where a pure stand of dense grass covers the ground, some people record the point strike with a check and a letter for the species, and then measure the distance to the nearest perennial of another species, which may be quite far.

3. **Record the habitat of the nearest perennial plant.** The water cycle influences plant habitat

and thus the kind of plants that can grow in an area. Indicate with a check whether the habitat is "Dry," "Middle," or "Wet" (xerophytic, mesophytic, or hydrophytic).

4. **Record the age of the nearest perennial.** Make a check mark in the appropriate column: "Seedling," "Young" (from seed or stolon), "Mature," "Decadent (Dying)," "Resprout" (i.e., not a new plant but coppice growth from an existing tree or shrub, or a rhizome from a parent grass plant).

5. **Record the form of the nearest perennial.** The form is the plant shape under the influence of grazing or rest. Your options are "Normal," where the plant appears vigorous and you see evidence of seed production, tillering (new stems at ground level) and branching, and a lack of old, stale growth; "Overrested"; "Overgrazed"; "Overbrowsed"; and "Dead" (due to overresting, overgrazing, or overbrowsing). If the plant form fits none of these descriptions, make a note of it in the "Other Comments" column.

6. **Note the species of the nearest perennial.** If you can identify the species, write its name in the column provided (very small seedlings are nearly impossible to identify). If you do not know the name, simply record "unknown."

When you have completed the sample points, subtotal the number of checks (or letters) in each column of each page. If your data are taken from 100-point hits, the combined subtotals (except "Distance to It") will automatically reflect percentage figures.

Summarizing the Data

Now is the time to separate the useful knowledge out of all the information you recorded on the monitoring data sheets. The process demands little computation but a good deal of thought. The monitoring summary form (fig. 3-20) breaks the process into sections that help synthesize information and bring it to bear on the status of the landscape you are attempting to create. Do the following:

1. Fill in the identification information at the top of the form. If your files ever become separated, there should be no doubt about which monitoring data sheets contributed to the summary.

2. If the sampling area was fairly uniform and you recorded 50 rather than 100 points, you will need to calculate percentages before you record the numbers on the summary form. Do this by multiplying the subtotaled figures by 2. For example, if 23 of the 50 dart points hit bare soil, the percentage of bare soil would be $23 \times 2 = 46$ percent.

3. Finally, transfer the following information from the monitoring data sheets:

   - *Plant name or species:* In the "Plant Name or Species" column on the left, list the species you were able to identify within each plant type (grass, sedge, forb, etc.), and to the right (under "No-%") how many of each (multiplied by 2 if you sampled only 50 points). Lump any species you weren't able to identify into one category ("Unknown") and record their numbers (multiplied by 2 if 50 points were sampled).

   - *Soil surface:* In the "Cover & Capping—%" column, record in each row the total of the relevant columns on the monitoring data sheets. In the "Evidence of—%" column, total the number of *Is*, *Ws*, *Bs*, *Ss*, and *Ls* that appear in the "Animal Sign" column on the monitoring data sheets and record in the appropriate rows. For the "Annuals" row, record the sum of the subtotals from the monitoring data sheets, and do the same for the "Soil Movement" row.

   - *Nearest perennial:* In the "Characteristics—%" column, enter the relevant numbers for the "Cool Season," "Warm Season," and "Year-Round Green Grass" rows. (Note:

If you were not able to identify every grass as cool- or warm-season or year-round green, the results here will not reflect a true percentage). Then enter the totals for rushes, forbs, shrubs, and trees.

   - *Average distance:* Calculate the total of the "Distance to It" column on the monitoring data sheets, divide that figure by 100 (or 50, if 50 points were sampled), and record the number.

   - *Age and Form—%:* Record the sum of the subtotals from the corresponding columns on the monitoring data sheets.

Analyzing the Data

Begin by filling in the identification information at the top of the monitoring form (fig. 3-21) to make sure this form is never confused with others. In the first year much of the data becomes baseline information. In subsequent years your recorded analysis becomes the basis of a great many decisions.

The questions asked on this form are the same questions asked in the basic monitoring procedure. If you alternate using the two procedures over the years, your analysis will be comparable year after year. The comprehensive monitoring analysis, however, takes into consideration not only your observations but also the actual data collected. Use the back of the form as needed.

Future Landscape Description

What are we trying to achieve in the area surrounding this transect? Record your answer in terms of the four ecosystem processes but also indicate whether there are specific landscape features you are attempting to create—such as brushy areas. Your answer to this question will likely remain the same year after year.

Progress Check

What progress have we made this year, compared to last? Here you rate the status of the four ecosystem processes according to the data included on the summary form. Note specific positive or adverse changes, or no change at all, in terms of community dynamics, water and mineral cycles, and energy flow.

COMMUNITY DYNAMICS

- *Cover:* This tells much about the successional level of the community. Obviously, an increase in bare ground and rock represents a decline. Litter, especially litter 2, is often a precondition for advancing succession. Basal cover is one indication of plant density.

- *Capping:* This aspect of soil surface condition also has profound implications for successional movement. Mature capping is usually a sign of stagnation or decline. Covered soil (except in sod-bound conditions) is usually a sign of advance. Immature, recent, and broken capping lies between the two but would be seen as an advance if these forms of capping have increased while mature capping has decreased.

- *Animal sign:* This information becomes more meaningful over the years as you watch it change. An "infestation" reflects a lack of balance and stability, but it is not necessarily bad. Certain species will predominate at given levels of succession or moments of time. Massive hatches of cicadas occur every seventeen years in eastern North America, for example, but cause little damage when predators are diverse and abundant.

- *Annuals present:* If you are attempting to return annual grassland to perennial grassland, the number of annuals should start to decrease relative to perennials.

- *Soil movement:* Erosion and bare ground are key indicators of noneffective water cycles, but low-successional communities and exposed, eroding soil go hand in hand.

- *Plant type and species:* Depending on the future landscape described in your holistic context, changes in the proportions here will show your progress toward the future condition required. Consider the general diversity, the proportion of annual plants, and the presence of high-successional plants. Changes in cool- and warm-season grasses are significant in terms of diversity.

- *Average distance:* This index of plant density (except in sod-bound conditions) is very important. Closely spaced plants hold soil and litter in place and keep soils covered. Baseline plant spacing information does not in itself tell you much, but comparisons to later monitoring will provide one of the most sensitive indicators of a trend. Expanding bare areas and falling plant density are signs of a declining ecosystem. Decreasing distances indicate the reverse.

- *Plant age and form:* Obviously, the mix and number of plants in the youngest class determine the community of the future. The proportion of moribund plants shows changes from the past. Plants that have been overgrazed, overbrowsed, or overrested may die prematurely or fail to reproduce, eventually causing a shift in succession.

WATER CYCLE

- *Cover:* Litter or basal cover enhances the water cycle. Excessive grazing of dry vegetation, or consumption of litter off the ground, may result in inadequate cover despite minimal overgrazing during the growing season.

- *Capping:* Capping is a primary indicator of a noneffective water cycle.

- *Animal sign:* Burrowing animals directly affect water cycle. Although more animal activity is usually better than less, evaluation is subjective.

- *Soil movement:* Soil movement is a direct indicator of a poor water cycle.

- *Plant type and species:* Plants with varying root depths get their water at different levels, so diversity reflects health. Think this through for your mix of plants. What conclusion can you make about depth of water table, transpiration, surface evaporation, and so on? Certain species associated with good or bad soil aeration (xerophytic = dry type; mesophytic = middle type; hydrophytic = wet type) may indicate changes in the water cycle. Cool- and warm-season grasses are not prime indicators of water cycle status. Are perennials increasing and improving the water cycle?

- *Average distance:* Changes in plant density may precede changes in litter retention and water cycle.

- *Plant age and form:* Water cycle changes may explain why certain seedlings do or do not survive, why classes of plants are dying, and other peculiarities of age distribution. Think it through. Overgrazed and overrested plants provide less litter to help cover soil in brittle environments.

MINERAL CYCLE

- *Cover:* Litter, especially litter 2, is a direct measure of improvement. Check how quickly litter seems to be breaking down.

- *Capping:* Capping generally inhibits effective mineral cycling.

- *Animal sign:* The decomposers and all creatures that live in the soil or burrow through it are agents of mineral cycling.

- *Soil movement:* Any loss of soil is a break in the mineral cycle.

- *Plant type and species:* Look for plants of varying root depth and note the general diversity of species. Are deep-rooted perennial grasses increasing?

- *Average distance:* Greater plant density implies more cycling and ability to keep soil covered and stable.

- *Plant age and form:* This is relevant as an indicator of growth activity. Obviously, a large proportion of dying plants, if decaying slowly, amount to a slowing of the cycle.

ENERGY FLOW

- *Cover:* Bare ground or rock obviously indicates the worst possible energy flow.

- *Capping:* Capped soil indicates less than possible organic matter and soil activity. Bare, capped soil always indicates a loss of potential energy flow.

- *Animal sign:* A better energy flow supports more activity.

- *Soil movement:* Since erosion usually means bare ground, it seriously affects energy flow.

- *Plant type and species:* Broad-leafed, middle-type plants indicate more energy flow. A good mix of both cool- and warm-season plants means a longer period for converting solar energy and thus a better flow. Plants that remain green year-round will be growing year-round and thus converting more energy. On land dominated by annual grasses, an increase in perennial grasses usually means an increase in energy flow.

- *Average distance:* This is an important indicator. Tighter density usually means that a greater volume of leaf is harvesting sunlight.

- *Plant age and form:* Remember that green leaf area exposed to sunlight determines energy flow. Overbrowsed, overgrazed, overrested, and dead or dying plants obviously do not convert maximum energy.

Influencing Factors

What natural or management factors might have influenced what you are seeing on the ground? Think in terms of natural forces—a fire or flood that swept through the transect area during the year; weather factors, such as a heavy downpour or a hailstorm that occurred in the area a few days before you made your observations, or a complete rainfall failure. If your stock have been in the paddock covered by the transect very recently, or not for months, that would be worth noting. If you created herd effect with an attractant, the land will be different than it would have been otherwise, and you should note that the herd effect occurred.

Change or No Change?

If adverse changes or no changes have occurred where change was planned, what is the underlying cause—what tools did you apply, and how have you applied them? If you identified a poor water cycle in question 2 because of pervasive capping and absence of litter, for example, you might focus on insufficient animal impact (partial rest) as

Box 3-6. Interpreting Your Results: Five Scenarios

Each of the following scenarios reflects the changes that have occurred since the previous monitoring one year ago, on brittle-environment properties whose managers seek a future landscape that includes healthy grassland, effective water and mineral cycles, and high energy flow.

Interpretations

1. Stock density is low (partial rest still too high), but there is some improvement.
2. Stocking rate is too high—all four ecosystem processes are worsening.
3. All is well; the land is regenerating, ecosystem processes are improving.
4. Stocking rate is too low and stock density is too low.
5. Animal impact is good, but the stocking rate is low. The manager is most likely rotating stock or not watching recovery periods carefully enough.

Think through each scenario and interpretation given to see if you can figure out the reasoning behind each interpretation. Then check your answers with those given in appendix 6.

Scenario	Bare Ground	Mature Capping	Litter	Broken	Plant Spacing	Over-grazing	Over-rest	Weeds or Woody Plants
1	Decrease	Decrease	Increase	Increase	No change	None	None	Increase
2	Increase	Decrease	Decrease	Increase	Decrease	None	None	No change
3	Decrease	Decrease	Increase	Increase	Decrease	None	None	Decrease
4	No change	No change	Increase	No change	No change	None	Increase	Increase
5	Decrease	Decrease	Increase	Increase	Decrease	Yes	Yes	Increase

the prime cause. (See box 3-6 for additional examples.) Think through each of the tools as you applied them during the year and how they tend to affect each of the four processes. One tool in particular—rest (partial or total)—is noteworthy because it tends to produce opposite effects in very brittle and nonbrittle environments.

The Tools and Their Effects

Here is a brief review of what each tool tends to produce in terms of community dynamics, water and mineral cycles, and energy flow (for a more detailed discussion, refer to *Holistic Management, third edition,* chapters 18 through 23 covering each tool).

Fire

- *Community dynamics:* Fire exposes soil and thus tends to inhibit the establishment of new plants that require litter, moisture, and low tempera-

ture fluctuations throughout the day and night. Growth of mature woody species is often stimulated by fire. Where one stem existed before, fire can lead to several more after a period in which the plant appeared to have been killed by the fire. Only a few woody species are killed by fire. In the short-term, fire tends to increase the diversity of species in grassland and woodland. Repeated fires often reduce diversity. Fire can produce mosaic patterns within a given area, creating an edge effect and thus a zone of greater biological diversity.

- *Water cycle:* Fire tends to reduce water cycle effectiveness because it exposes soil and destroys litter. The lower the rainfall and the more frequent the fire, the greater this tendency.

- *Mineral cycle:* Fire tends to speed mineral cycling in the short term but if used repeatedly tends to slow mineral cycling in the long run. The drier the area and the more frequent the fire, the greater this tendency.

- *Energy flow:* In the short-term, fire tends to produce a decrease in oxidizing, overrested plants, but, because the soil exposure leads to less effective mineral and water cycles and changes in the plant community, fire could reduce energy flow in the long term. The drier the area and the more frequent the fire, the greater this tendency.

Technology

The thousands of technologies that fall under this heading cannot easily be broken down into categories on which to base general tendencies. Use your common sense. You know that the use of technologies such as pesticides, for instance, will tend to reduce biodiversity and thus adversely affect all four ecosystem processes. The context checks will help ensure that a decision to use them anyway is a conscious one.

Rest (Partial or Total)—Nonbrittle Environments

Rest is the most powerful tool we have to restore or maintain biodiversity and soil cover in completely nonbrittle environments.

- *Community dynamics:* Biological communities develop to levels of great diversity and stability.
- *Water and mineral cycles* build and maintain high levels of effectiveness.
- *Energy flow* reaches a high level.

Rest (Partial or Total)—Very Brittle Environments

In very brittle environments, rest in either form—partial or total—applied continuously tends to damage biodiversity and soil cover. (See the notes under "Animal Impact," which follows, for a more detailed discussion on the tendencies of low animal impact, the flip side of partial rest).

- *Community dynamics:* Rapid biological decay, especially in grasses, gives way to gradual chemical oxidation and physical weathering. Con-

sequently, biological communities decline and greater simplicity and instability ensue, particularly in grasslands and savannas. The lower the rainfall, the greater the adverse effect. In high rainfall areas, grasslands tend to be replaced by woodland.

- *Water and mineral cycles* become less effective.
- *Energy flow* declines significantly.

Living Organisms

When living organisms other than large, grazing animals (covered below) are used in the management of ecosystem processes, you must analyze each specific case. We are not yet aware of any general tendencies that could guide land managers. Because there is a danger of organisms used to eradicate insect or plant populations subsequently finding other hosts, extreme caution is needed, particularly in the case of genetically engineered organisms released for such purposes. In some cases, living organisms have been used to eradicate populations that might have been dealt with more effectively through the use of grazing and high animal impact. For example, insects have been released to deal with problem plants, such as moths to help reduce prickly pear, when the offending, dominating plants were a problem due to partial rest.

Animal Impact (High)

In environments tending toward the brittle end of the scale, the most common problem in the early days is partial rest that goes unrecognized. Consider reducing the number of herds or even herding (on large ranches) because increasing the number of paddocks in such situations does little to overcome partial rest.

- *Community dynamics:* Periodic high animal impact tends to promote the advancement of biological communities on bare, gullied, and eroding ground. In dense grassland, high impact tends to maintain the biological community at the grassland level, preventing a shift to woody communities in both brittle and nonbrittle environments.
- *Water and mineral cycles:* Periodic high animal impact generally improves water and mineral

cycles. Since they tend to sustain grasslands in nonbrittle environments, water and mineral cycling *may* not be as effective as they would be if the community advanced to forest.

- *Energy flow:* Because periodic high impact tends to build community complexity and improve water and mineral cycles, energy flow tends to improve as a direct consequence. However, when used to maintain grassland in lieu of forest, energy flow in nonbrittle environments, or in brittle tropical areas that can support a solid woodland canopy, will not reach its full potential.

Animal Impact (Low)

- *Community dynamics:* Low animal impact, or partial rest, has little impact on nonbrittle grasslands that cannot advance to forest because of some natural factor, such as shallow soil or elevation; otherwise, the tendency is to move from grassland to forest. In brittle environments, however, low animal impact tends to produce bare ground, as it disturbs algae–lichen communities but does not stimulate the establishment of more complex communities. It allows plant spacings to increase and on a larger scale generally has effects remarkably similar to those of total rest. Under low impact, dense grassland with close plant spacings may proceed toward woody communities and forbs, but those will give way to a landscape of scattered shrubs or trees and much bare, or algae- or lichen-covered ground unless rainfall is sufficient to sustain a full woody cover.

- *Water and mineral cycles:* In brittle environments, low animal impact reduces mineral and water cycles below the land's potential. In nonbrittle environments, it has little effect.

- *Energy flow:* Low animal impact generally reduces energy flow below its potential in brittle environments but can increase it in nonbrittle environments if there is movement from grassland to forest.

Grazing

- *Community dynamics:* Grazing tends to maintain grass root vigor, soil life, and structure and retard shifts toward woody or herbaceous species.

- *Water and mineral cycles:* Grazing enhances both of these cycles by maintaining healthier and more stable root mass, increasing microorganism activity and aeration, and producing plants with more shoots and leaves, providing much needed litter in brittle environments.

- *Energy flow:* Grazing increases energy flow both above and below ground. It does this in brittle environments by preventing old, oxidizing blockages of material and in both environments by promoting vigorous root and leaf growth. Healthier, more massive root systems also support millions of microorganisms and other life underground.

Overgrazing

- *Community dynamics:* In brittle environments, overgrazing reduces litter and soil cover, damages grass roots, and fosters shifts away from grassland toward herbaceous forbs ("weeds"). In nonbrittle environments, a solid mat of grass tends to occur. In both environments, soil-enhancing legumes simply disappear.

- *Water and mineral cycles:* Overgrazing reduces water and mineral cycling by exposing soil and limiting the production of potential litter (important for brittle environments) and leads to grass root reduction, which in turn fosters soil compaction in both brittle and nonbrittle environments.

- *Energy flow:* Overgrazing cuts energy flow because it reduces plant roots and exposes the soil surface. However, in nonbrittle environments, where it produces a shift to woody communities the climate can sustain, it would eventually lead to increased energy flow.

Proposed Actions

What are we going to change in this next year to keep our land moving toward the landscape described in our holistic context? If your monitoring indicates that you are veering off track you need to take action right away. In most cases, this will require the use of a tool other than the one that led to the adverse change, or a modification in how you applied the tool. You have identified the conditions and the causes and therefore should now be able to identify the tool required to bring about a remedy. Determine which management guidelines apply and then decide how you will use that tool. When you have completed this step, you will find that you have outlined the actions you should take. Run those actions through the context checks to ensure they are aligned with your holistic context and modify your plans accordingly. *Holistic Management* (third edition) explains the management guidelines and context checks in detail.

Using the example of a poor water cycle again, for which insufficient animal impact could be the cause, say you discover that two management guidelines—stock density and herd effect—apply most directly. *Holistic Management* will tell you various ways to increase stock density and herd effect, and then you will have to run each option through the context checks. To improve density, you could consider increased fencing, amalgamation of herds, or increased stocking rate. The context checks will help you decide which. To improve herd effect, you could train animals to respond to attractants of various kinds, or strip graze at ultra-high density using temporary fencing (to produce the behavior change that results in herd effect), or simply advance to herding. Record the changes you plan to make.

Once you have completed each of your summary and analysis forms, make sure you file them together with their data forms and photos in a way that prevents them from being separated.

A Note on Shortcuts

Make every effort to monitor as thoroughly as possible, whether you are using the basic or the comprehensive procedure, especially in the first years. If for any reason you cannot complete the whole process in a given year, the following data are most important:

- Fixed-point photos
- Soil surface information
- Thorough notes on observations (questions 1–5 on the monitoring analysis forms)

Summary

IF GRAZING PLANNING is the route to managing your piece of the natural world, monitoring is the best way to learn from it. Monitoring a farm or ranch involves both a constant attitude of openness and curiosity and a self-disciplined labor of measuring, recording, and photographing actual data.

Anyone who has ever contemplated the life of peasants and indigenous people is amazed at their enormous pool of wisdom and lore and has wondered how they got it. What unfortunate individual discovered—and alerted his survivors to—the toxic powers of the fly amanita mushroom? Why does an ancient Greek manuscript declare that blowfish crawl out of the water and mate with goats at certain phases of the moon? And why does that idea still persist among oldtimers on the U.S. Gulf Coast?

In the mushroom case—monitoring. In the blowfish lore we suspect not.

You've got to monitor to understand the difference between myth and substance. And you've got to try to understand everything in order to do anything. You won't get anywhere standing on the top of your hill or sitting behind your desk telling your animals and crops what to do, forgetting for the moment about the thousands of other rebellious and independent creeping, burrowing, flying, thrusting, and twining things that surround you.

You have to hark to all of them as well as forces like wind, water, and sun that you never expect to pay attention to you.

And you have to record what you learn so you can think about what it means, remember it next year, and pass it on in a comprehensible form to heirs, hands, and others—and most of all so you can use it to keep your planning vital and flexible and get better at what you do.

Don't whine when the range goes bare in January and nothing grows till June. Monitor the grass. Don't weep when starving elk bust your fences and plunder your hay. Monitor their winter range. Don't protest when your topsoil leaves for the ocean. Monitor ground cover. When old Aunt Maude left the world muttering on her deathbed that there never was a family *bundküchen* recipe except to add this and that until it smelled right, she was telling you to . . . monitor.

PART 4

HOLISTIC LAND PLANNING

PART 4

HOLISTIC LAND PLANNING
Designing the Ideal Layout of Facilities for a Grazing Operation

Holistic Land Planning and Holistic Planned Grazing developed in tandem as both grew out of our new understanding about the role of time in grazing and trampling. To keep large herds of animals moving continually requires new thinking about the way fencing, water points, roads, and handling facilities are laid out—especially when you're dealing with large tracts of land that require a considerable investment in that sort of infrastructure. Once we no longer were constrained to fencing based on soil types, many possibilities opened up for laying out the needed infrastructure in ways that could improve animal handling, reduce capital and running costs, and increase the profitability of ranching operations.

Because Holistic Land Planning is basically a procedure for planning the layout of infrastructure for livestock grazing, the term *land planning* is something of a misnomer. This is what most ranchers have preferred to call it, so we've kept the name.

This procedure is not recommended for smaller farms in less brittle environments, or where grazing animals aren't the main line of production. In those cases, farmers might consider the many fencing layouts shown in the next section ("Creating Your Plan"), but because their land area is relatively small, distance to water for livestock is minimal, and quite often simple electric fencing can be used and layouts easily changed, there isn't a need for a planning procedure as elaborate as this one.

Some may ask why planning the layout of infrastructure comes last in the sequence of planning procedures. The reason is both practical and psychological. You cannot in practice assess the possibilities for your land plan without a background of financial and grazing planning. More importantly, many ranchers think Holistic Management cannot begin until they've fenced their land in some special way. In fact, whether or not you ever invest anything in fencing, *you must start managing the land as you find it.* Start planning your grazing using the fencing you have. Producing the ideal long-term plan for fencing, watering, and handling livestock is not a starting point. It is an end point toward which you build based on your grazing and financial planning experience.

Conceiving the matter backward often leads to quick disaster, yet no end of impatient people rush to borrow heavily for fences, water, or whatever, basing their plans on existing infrastructure, and creating a layout that proves inconvenient and costly in the long run.

A holistic land plan represents the marriage of a holistic context and step-by-step practice. Its implementation will proceed, like biological succession, as one stage makes the next one possible. Because investments in infrastructure represent long-term commitments, this land-planning procedure and its gradual implementation acquire extreme importance.

Alongside Holistic Financial Planning and Grazing Planning, Holistic Land Planning is a major aspect of management for the rancher. It differs, however, in that it can involve very long term and occasionally irreversible commitments. Success depends on exploring all the possibilities and trying to avoid decisions that permanently eliminate large areas of choice.

The steps for developing a land plan emphasize creativity and broadness of vision. The plan you create will map all future developments for effectively managing larger herds of livestock on your land, and it will affect future generations as well. The land plan you develop will be implemented over time in steps controlled by your holistic financial plan. Each year, for instance, planned fences or water points will be put in, but only those that "earn," rather than cost, money. And each year your grazing plans will ensure the best possible production from your animals and land with the infrastructure you currently have.

Anyone can whip out a "plan" without following the steps described in this section, but no abstract blueprint for a distant future will have as much immediate practical use as the knowledge you'll gain from completing the tasks in each of these steps. In fact, your final plan will develop almost as a by-product of the process. These are the steps:

1. Gather the information (factors affecting your plan, grazing unit and herd sizes, infrastructure needed).
2. Prepare maps and overlays.
3. Decide grazing unit and herd sizes.
4. Generate a series of plans.
5. Design the ideal plan.
6. Implement your plan.

The first section of part 4, "Mastering the Basics," will cover the first three steps. The second, "Creating Your Plan," covers the last three steps as well as potential layouts and hardware.

Mastering the Basics

Land planning, to be successful, should never be a solitary exercise. Your land plan will affect more than you, the owner or manager. It will also affect those who work with you, your neighbors, and the generations to come. Bring as many minds to the table as you can, both when gathering the information that goes into the plan and in brainstorming possible plans later.

If you're ranching on public lands, or sharing your land with multiple users, you will need to enlarge participation even further. If you don't, the fences or water points or other structures you plan could easily lead to unnecessary conflict.

Well-informed communication among those contributing to the planning is essential. And for that, you will need good information and good maps. Lacking this, good discussions tend to turn into arguments. Spare no pains in gathering every scrap of background material that will help you plan wisely.

Gathering Important Information

A land plan showing the ideal layout of the infrastructure you will need for running large herds of livestock is no better than the information that goes into creating it. Seek ideas from everyone who has an interest in your land or knowledge to contribute. People who participate in lending their ideas become supporters, even if they say no more than "You have a rare plant over there," or "We know that most deer fawning happens in this area," or "Sage grouse breed in that draw."

Start a Checklist of Issues

For each item on the list, you want to answer questions such as: Where on the land do we need to consider this? Who is involved? What would I like to see happen? Show the list to others and get their input. Keep adding to the list. Think in terms of the following categories:

- **Natural issues:** Weather factors, such as the direction of prevailing winds and areas that are likely to be covered in deep snow or prone to flooding; geographic features such as water sources and riparian areas, eroding catchment areas that lead onto your property, major differences in soil types (because they could determine the placement of crop fields); areas of major fire threat and wildlife concerns such as the need for roosting, mating, and nesting sites and seasonal movement routes; areas of heavy predation and areas where endangered species are present.

- **Social issues:** Present ownership boundaries (some properties may be divided among different owners or portions may be leased) and future boundaries (an estate plan might require that the land be divided later among several family members), areas where mineral rights are leased or could be, water rights (both yours and your neighbors') and how those might affect your own water use, any future developments planned on the land surrounding the property, recreational access areas for hunters or hikers, areas prone to vandalism, archaeological sites needing protection, and so on.

All these issues can affect your plan in one way or another. The prevailing winds might influence the positioning of livestock movement corridors that can serve a dual purpose as firebreaks. Locating a fence where it would inhibit the movement of the public in an area they have access to could invite vandalism.

Identify Management Factors

Envision your future management needs and desires, not just those related to profit but taking into account aesthetic concerns, wildlife, possible crop fields, and other potential uses of the land. Consider the following, as well as the factors noted in box 4-1:

- Livestock production
  - *Possible future stocking rates:* These will influence the size and number of grazing units, paddocks, and handling facilities.
  - *Herd size:* This will enable you to determine how much water needs to be available in any one place. In the future, you are likely to require many more paddocks (subdivided with permanent fence, or temporary fence for strip grazing) than you now have. But fewer paddocks if livestock are to be herded.
  - *Access to water and other points:* These could include a milking facility or overnight holding area. You want to avoid repetitive movement over the same ground, and to site corridors and any other sacrificial areas carefully to minimize damage from trampling and pollution problems.
  - *The brittleness scale:* Less brittle environments tend to need high stock density—meaning small paddocks—and frequent livestock moves to ensure good animal performance. More brittle environments, and consequently larger ranches, tend to need higher herd effect, as opposed to stock density.

- *Water storage:* Consider potential dam sites (check your water rights) if additional water is going to be needed, and where to site bulk water storage tanks—both largely determined by topography.
- *Multiple species:* Think seriously about running more than one species of livestock. In many situations doing that will result in better use of a wider range of forages and higher production. If you are now running cattle only, consider adding sheep or goats, hogs, pastured poultry, and so on—all of which could affect the amount and type of fencing needed, water availability, and handling facilities. Though you may not run additional species for years to come, planning for the possibility now saves having to modify your plan later, when it would be more costly to do so.

- **Crop production:** Will you run animals on crop fields at certain times of the year? If you irrigate, do you plan to change the form of delivery? How would that affect your historical use (there could be legal implications)? Will your machinery have access? Remember: Small fields are more manageable ecologically than large ones; the more "edge" (where different habitats such as fields, hedgerows, tree belts meet) you have, the more you encourage diversity in insect, bird, and game populations.

- **Timber production:** Will you graze livestock in the forest area to assist with fire mitigation or for any other reason? You will need to ensure access to the timbered area and try to minimize soil damage from machinery. What harvesting methods will be used (clear-cut, in which large swaths of forest are harvested at once; selective harvesting, in which the best trees are taken; or selection harvesting, in which the best trees are left to produce seed for future generations)?

Start a List of the Infrastructure Needed

Based on all the management factors you have considered, you should be starting to get an idea of the infra-

Box 4-1. Breadth and Depth in Your Thinking

You'll begin to understand why gathering information must precede drawing lines on maps or overlays as you confront decisions that can influence everything that follows, such as whether to hire herders or build fences, where livestock can be watered or handled, or how you lay out crop fields.

Herding versus Fencing

Are there public opponents to fences? Are there legal restrictions on public or leased lands? Can herding ease predator problems? Does permanent fencing fail the context checks? Will movable fencing or permanent posts with movable tape or wire need to be used?

Herding requires a good deal of skill and initial training of animals. What about labor costs and availability? Can you substitute dogs for some of the people to do the herding work? What do you know about stock dogs, and where can you learn more? How about the economic questions? How large would your herd size have to be to produce the income to cover herder salaries? How does the cost of herders, who can create high animal impact consistently through herd effect, compare to

the cost of fencing, which can only achieve high animal impact through stock density unless fencing is pushed to extreme levels and animals moved every few hours?

Are there partial options involving seasonal herding on some land or skillful use of natural barriers, drift fences, movable fences, polywire, and so forth in combination with herding?

Cropland

Now, and probably more in the future, you may want to integrate crop and livestock production to keep soils covered, to regenerate them, to increase biodiversity, or any number of reasons.

Aside from the usual advantage of some grazing and hay production, your information gathering may turn up ways to apply animal impact, rather than fire or machinery, to deal with crop residues, or strategic feeding of winter hay, or other techniques.

What you discover about the relationship of field size and crop sequence to weed and pest control and wildlife may well lead to a fence layout that lets you easily incorporate crop fields into the grazing plan.

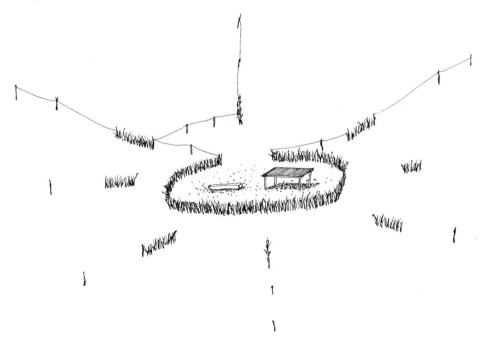

Box 4-1. In herding situations, informal fences extending some distance from the center can help direct traffic where a big sacrifice zone would otherwise develop.

structure required. Draw up a list of the basic facilities, and keep adding to it as you move through the next steps.

Circulate Your Lists and Ask for Feedback

When you have written down people's opinions and suggestions on the issues, management factors, and infrastructure desired, and the lists have circulated long enough to ferment a bit, you can begin to consider specifics.

Preparing Maps and Overlays

A good topographical map, one that shows contours as well as natural features, such as rivers, is essential for land planning. It should not show existing fences, water points, ranch roads, and the like, lest they bias your thinking toward the old facilities and stifle creativity. You will put those features on an overlay to use later when considering new ideas.

Create a Master Map of the Property

You will need the best contour maps available and of a scale suitable to work on in some detail. In the United States, U.S. Geological Survey 1:24,000-scale maps are generally adequate for planning most ranches and other extensive properties, though you may have to cut and paste several of them together to create your master map. The only constructed features that should appear on this map are those that would be illegal or genuinely impractical to change, such as public roads, railway tracks, homesteads, or villages.

Map Your Future Landscape

If your land covers an extensive area and varied terrain, and you intend to produce major changes, record on a map, or transparent overlay, the essential features of the future landscape you intend to produce—where croplands will be, heavily wooded or brushy areas, wetlands, or open grassland areas, and so on. (You will use this map when later selecting the best of many possible infrastructure layouts.)

Prepare a Map of Existing Developments

Existing developments would include water points, buildings, fences, croplands, major roads, working facilities,

and so on. Prepare a map that shows all these things, and then also set that map aside for later use.

Create Overlays

The information you gathered in the first step now needs to be recorded on clear plastic or transparent paper sheets that will be superimposed on the master map. You can get several kinds of clear plastic overlay material in rolls or sheets, as well as indelible and wipe-off markers in a rainbow of colors, from an engineering supply house, art store, or blueprinter.

Several items can be put on one overlay as long as the sheet doesn't become cluttered. Usually, the information will fall into three or four categories, requiring three or four transparent overlays. Some of the following examples could be grouped together; others will require individual sheets:

- Map of your future landscape (if you have mapped the changes you intend to produce)
- Hunting areas
- Winter wildlife range
- Wildlife factors, such as roosting, mating, and nesting sites; food plants; movement routes
- Deeds, leases, and permits
- Rights-of-way
- Features involving multiple users (camp sites, trails, etc.)
- Fire danger and prevailing wind
- Crop or timber areas
- Inaccessible areas
- Existing facilities, fences, watering facilities, and roads
- Bulk water storage sites
- Snowdrift areas
- Flood areas
- Estate plans (may involve future land divisions)
- Water rights and leases
- Mineral rights and leases

Study the maps in figures 4-1 and 4-2. Note how the fence pattern on the second map immediately suggests a grazing unit layout, even though in this case it derives

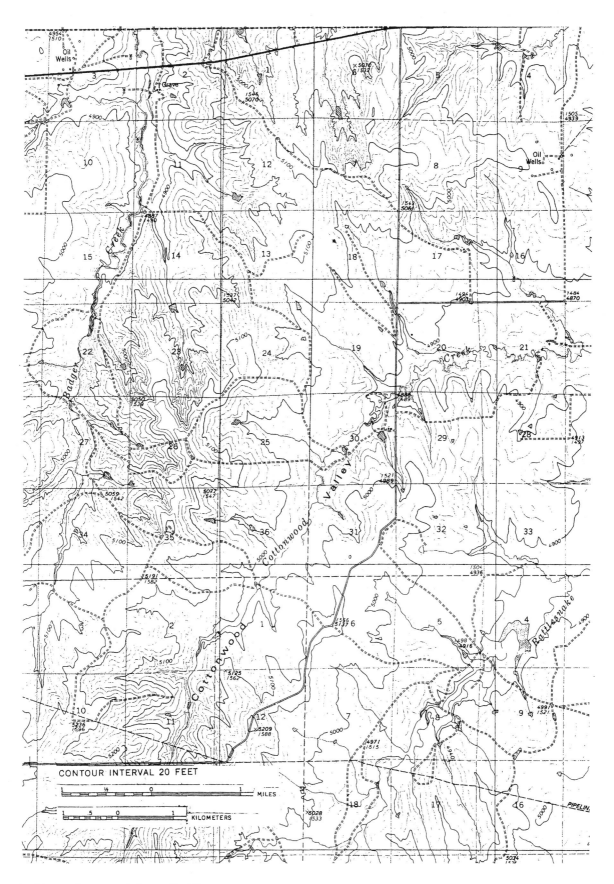

Figure 4-1. Topographic map including no fixed features.

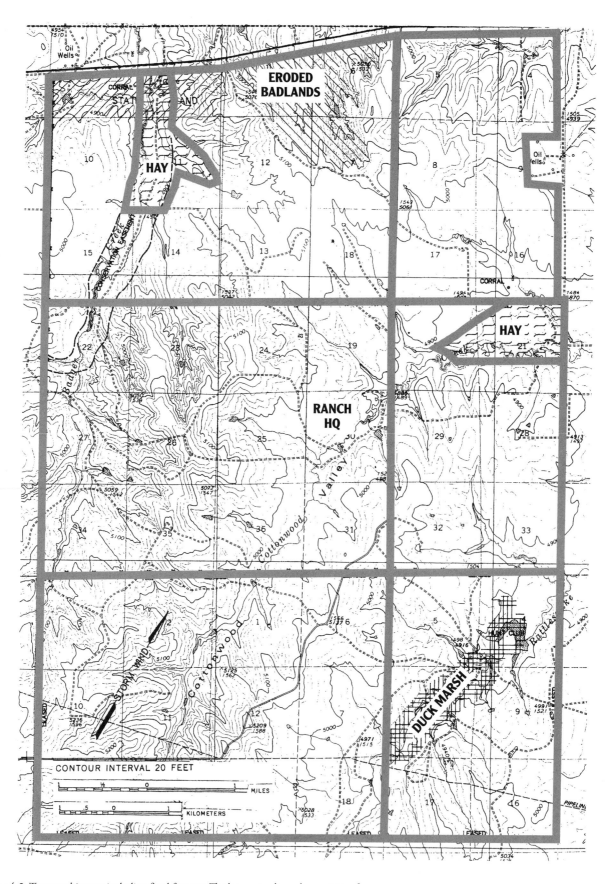

Figure 4-2. Topographic map including fixed features. The heavy gray lines show existing fences.

from old property boundaries that have no relationship to the lay of the land. This is why you should do your creative planning on a map stripped down to basic geography.

Measure Acres and Hectares

In land planning, you must constantly relate land area to stocking rates, grazing unit sizes, and, of course, topography. Unfortunately, most people are terrible judges of area. Though an acre happens to be about the size of an American football field, who can easily tell how many football fields fit into an irregular 12,000-acre grazing unit with a river and a highway through the middle? Google Earth Pro and numerous other software applications (many of them free) can help you calculate land area, and are fast and easy to use. If you don't have access to them, you can calculate the area (in acres or hectares) by hand using the following technique.

Acres

Squares of overlay material or even thin paper divided into ten-acre squares make these measurements easy on a map. The grid shown in figure 4-3 is for the 1:24,000-scale U.S. Geological Survey "Seven Minute Series" map often used in planning. Remember: You're not buying urban lots by the square foot. A quick and dirty estimate by grid does fine.

Among other things, such grids show that there is some logic after all in the American system of measurement. One-eighth mile squared is ten acres, and 1¼ miles squared is 1,000 acres. You can take both of these units right off the mile scale of any map.

Hectares

A one-centimeter grid works well for those using the metric system. Maps covering extensive land areas tend to be scaled to at least 1:50,000 and often much greater—1:100,000 or more. The clue to the area in a grid square is the scale itself. On a 1:10,000–scale map, which is typically used in more intensive agricultural areas, each centimeter represents 100 meters.

Each one- by one-centimeter grid square is one hectare (100 meters by 100 meters = 10,000 square meters = 1 hectare). On a 1:50,000-scale map, each centimeter represents 500 meters, so a one- by one-centimeter grid square is twenty-five hectares (500 meters by 500 meters). Simply multiply the number of grid squares that fully cover the block of land being measured by the area per grid square for your map scale. Given that you are not looking for decimal-point accuracy in extensive land situations, it is relatively easy to estimate the area of a grid square that covers only a portion of the block. For example, at 1:50,000, if a grid square covers only thirty-three percent of a paddock's area, assess the grid square as eight hectares, and add that to the sum of all the fully covered grid squares.

Deciding Grazing Unit and Herd Sizes

You'll need a rough idea of how large the grazing units will be and how many animals you might run in them to help you make better decisions about number and size of paddocks, water developments, and the like. We use the term *herd* to refer to any group of animals run together, including flocks of sheep, or mixed herds of cattle, sheep, goats, and so on. Remember that a grazing unit is the area planned on a single grazing chart and is not a particular shape or fencing layout.

On smaller properties, you might require only one grazing unit, though livestock management requirements may call for more, at least initially. As you gain experience and come to realize the true cost of running more than one herd, you may well combine two grazing units into one for the whole property. Because smaller properties often have a higher carrying capacity, however, herd size might still be large and become a critical consideration.

As in many other aspects of management, you can't find precise answers through rigorous mathematics because too many subjective factors enter in. But to get your thinking on the right track, consider your land as flat and uniform, with water available anywhere (if you're willing to dig for it).

Where such conditions actually exist, a good argument can be made for a "wagon wheel" fencing layout, with water at the center and fences radiating at precise intervals. The radial design has many advantages, especially in large, dry areas. Once you let go of the idea that the paddocks must be equal and symmetrical and actually look like a wagon wheel, you can shape them to fit rough and hilly country and still have them feed to a common center.

Measuring Acreage Using a Grid and Topographic Map

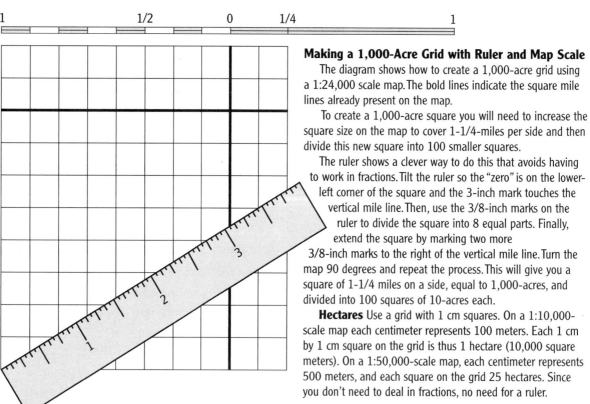

Making a 1,000-Acre Grid with Ruler and Map Scale

The diagram shows how to create a 1,000-acre grid using a 1:24,000 scale map. The bold lines indicate the square mile lines already present on the map.

To create a 1,000-acre square you will need to increase the square size on the map to cover 1-1/4-miles per side and then divide this new square into 100 smaller squares.

The ruler shows a clever way to do this that avoids having to work in fractions. Tilt the ruler so the "zero" is on the lower-left corner of the square and the 3-inch mark touches the vertical mile line. Then, use the 3/8-inch marks on the ruler to divide the square into 8 equal parts. Finally, extend the square by marking two more 3/8-inch marks to the right of the vertical mile line. Turn the map 90 degrees and repeat the process. This will give you a square of 1-1/4 miles on a side, equal to 1,000-acres, and divided into 100 squares of 10-acres each.

Hectares Use a grid with 1 cm squares. On a 1:10,000-scale map each centimeter represents 100 meters. Each 1 cm by 1 cm square on the grid is thus 1 hectare (10,000 square meters). On a 1:50,000-scale map, each centimeter represents 500 meters, and each square on the grid 25 hectares. Since you don't need to deal in fractions, no need for a ruler.

Figure 4-3. Squares of overlay material or transparent paper divided into ten-acre squares (or one-hectare or twenty-five-hectare squares, depending on map scale) make measuring land area on a map easy.

Whether you vary the design or build something quite different, you need to start thinking in terms of area per herd, location of handling facilities, and distance to water. A circle or square representing the size of an optimum grazing unit cut to scale from paper or overlay material, as shown in figure 4-4, will enable you to see on the map how these relationships (distance to water, etc.) apply to your land.

Your final decisions may bear no resemblance to these planning circles, and in any case no grazing unit is an immutable structure. We define a grazing unit only as a piece of subdivided land, planned as one unit. In practice, therefore, you can treat any combination of paddocks associated with any number of handling and water facilities as a grazing unit. Sometimes the whole property might be planned as one grazing unit; at other times you might plan the same paddocks as two or more grazing units, depending on the herds you run—the point being that a grazing unit is what is planned on a grazing chart as a unit at a certain point in time, rather than any particular layout of paddocks. Nevertheless, keeping in mind the chunks of land you'll probably plan as a grazing unit most of the time will greatly focus your thinking.

The factors to consider are the following:

- *Stocking rate:* If you set up a grazing unit big enough for your optimum herd now, and better management and improving land doubles or triples carrying capacity, you may have to run more animals at one time than you can gracefully manage just to make efficient use of forage. (See fig. 4-5, "Grazing Cells and Herd Sizes," and box 4-2, "Thinking Through a Grazing Unit Size/Herd Size Problem.")

- *Herd size:* Cattle, sheep, and goats can thrive in herds of any size. No limit has been found. Nevertheless, labor, handling facilities, water, and management factors such as calving, lambing, and kidding or weaning will offer significant practical challenges, and herd size may become unwieldy after a certain point, which you must determine for yourself.

- *Number of herds:* Although a single herd is ideal, there will likely be times when two or more herds might be run.

- *Paddocks per herd:* Ultimately, you will generally want many paddocks per herd to enable you to achieve great versatility and a consistently high graze-and-trample-to-recovery ratio for best performance of both land and livestock. So in laying out probable grazing units, it is important to think of their size and shape so they can be subsequently divided into many paddocks without needing realignment. (If you plan to herd without fencing, far fewer paddocks are required because herders will graze smaller areas within a paddock for a few days at a time until the whole paddock has been grazed for its allotted time.)

- *Water supply:* Although cattle in Hawaii can survive on dew and the moisture contained in plants, water is frequently a limiting factor elsewhere. Fifteen gallons per day per standard animal unit (SAU) (or forty to eighty liters/large stock and five to fifteen liters/small stock) is a common requirement. Given your estimates of future herd size, stocking rate possibilities, and grazing unit size, you can compute the bulk delivery of water required. At times, a single well (or borehole) may not provide all you need, so plan for the possibility that you may eventually have to pipe water from more than one source. Planning for the possibility now could avoid heavy expenditures later.

- *Distance to water:* Grazing unit size will also be limited by how far your animals must travel to water. This can vary greatly. In mountainous country, for instance, livestock don't necessarily come in to drink daily. In areas of Africa and inland Australia, it is routine for animals to graze five miles (eight kilometers) and more to and from water. You will need to plan for what you feel is best in your situation (see figs. 4-4 and 4-5).

Making Planning Circles

Acres ÷ 640 acres/square mile = square miles

$\sqrt{\text{square miles}}$ = side of square grazing unit

$\sqrt{\dfrac{\text{square miles}}{3.14}}$ = radius of a circular grazing unit

Miles x 63,360 inches/mile ÷ map scale = inches on map

Hectares ÷ 100 hectares/square kilometer = square kilometers

$\sqrt{\text{square kilometers}}$ = one side of square grazing unit

$\sqrt{\dfrac{\text{square kilometers}}{3.14}}$ = radius of a circular grazing unit

Kilometers x 10,000 ÷ map scale = centimeters on map

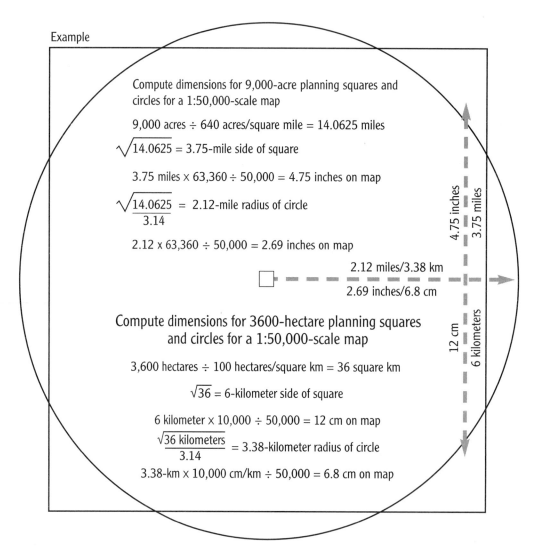

Example

Compute dimensions for 9,000-acre planning squares and circles for a 1:50,000-scale map

9,000 acres ÷ 640 acres/square mile = 14.0625 miles

$\sqrt{14.0625}$ = 3.75-mile side of square

3.75 miles x 63,360 ÷ 50,000 = 4.75 inches on map

$\sqrt{\dfrac{14.0625}{3.14}}$ = 2.12-mile radius of circle

2.12 x 63,360 ÷ 50,000 = 2.69 inches on map

2.12 miles/3.38 km

2.69 inches/6.8 cm

4.75 inches / 3.75 miles

12 cm / 6 kilometers

Compute dimensions for 3600-hectare planning squares and circles for a 1:50,000-scale map

3,600 hectares ÷ 100 hectares/square km = 36 square km

$\sqrt{36}$ = 6-kilometer side of square

6 kilometer x 10,000 ÷ 50,000 = 12 cm on map

$\dfrac{\sqrt{36 \text{ kilometers}}}{3.14}$ = 3.38-kilometer radius of circle

3.38-km x 10,000 cm/km ÷ 50,000 = 6.8 cm on map

Figure 4-4. Circles or squares representing optimum grazing unit sizes, cut to scale from transparent paper or overlay material, greatly assist in land planning. They allow you to consider a number of possible grazing unit sites, simply by moving them around on the map. Even though you will ultimately design grazing units that are shaped quite differently, the thinking that went into figuring the size of the planning circles or squares—area per herd, distance to water and handling facilities—will be critical in determining the final design.

Grazing Cells and Herd Sizes

Possible Cell Sizes (acres or hectares)

Stocking Rates	500	600	700	800	900	1000	2000	3000	4000	5000	6000	7000	8000	9000	10,000	11,000	12,000	13,000	14,000	15,000	16,000	17,000	18,000	19,000	20,000	21,000	22,000	23,000	24,000	25,000
1:1	500	600	700	800	900	1000	2000																							
1:2	250	300	350	400	450	500	1000	1500	2000																					
1:3	167	200	233	267	300	333	667	1000	1333	1667	2000																			
1:4	125	150	175	200	225	250	500	750	1000	1250	1500	1750	2000																	
1:5	100	120	140	160	180	200	400	600	800	1000	1200	1400	1600	1800	2000															
1:6	83	100	117	133	150	167	333	500	667	833	1000	1167	1333	1500	1667	1833	2000													
1:7	71	86	100	114	129	143	286	429	571	714	857	1000	1143	1286	1429	1571	1714	1857	2000											
1:8	63	75	88	100	113	125	250	375	500	625	750	875	1000	1125	1250	1375	1500	1625	1750	1875	2000									
1:9	56	67	78	89	100	111	222	333	444	556	667	778	889	1000	1111	1222	1333	1444	1556	1667	1778	1889	2000							
1:10	50	60	70	80	90	100	200	300	400	500	600	700	800	900	1000	1100	1200	1300	1400	1500	1600	1700	1800	1900	2000					
1:15		40	47	53	60	67	133	200	267	333	400	467	533	600	667	733	800	867	933	1000	1067	1133	1200	1267	1333	1400	1467	1533	1600	1667
1:20				40	45	50	100	150	200	250	300	350	400	450	500	550	600	650	700	750	800	850	900	950	1000	1050	1100	1150	1200	1250
1:25							80	120	160	200	240	280	320	360	400	440	480	520	560	600	640	680	720	760	800	840	880	920	960	1000
1:30							67	100	133	167	200	233	267	300	333	367	400	433	467	500	533	567	600	633	667	700	733	767	800	833
1:35							57	86	114	143	171	200	229	257	286	314	343	371	400	429	457	486	514	543	571	600	629	657	686	714
1:40							50	75	100	125	150	175	200	225	250	275	300	325	350	375	400	425	450	475	500	525	550	575	600	625
1:45								67	89	111	133	156	178	200	222	244	267	289	311	333	356	378	400	422	444	467	489	511	533	556
1:50								60	80	100	120	140	160	180	200	220	240	260	280	300	320	340	360	380	400	420	440	460	480	500
1:55								55	73	91	109	127	145	164	182	200	216	236	255	273	291	309	327	345	364	382	400	418	436	455
1:60								50	67	83	100	117	133	150	167	183	200	217	233	250	267	283	300	317	333	350	367	383	400	417
1:65									62	77	92	108	123	138	154	169	185	200	215	231	246	262	277	292	308	323	338	354	369	385
1:70									57	71	86	100	114	129	143	157	171	186	200	214	229	243	257	271	286	300	314	329	343	357
1:75									53	67	80	93	107	120	133	147	160	173	187	200	213	227	240	253	267	280	293	307	320	333
1:80									50	63	75	88	100	113	125	138	150	163	175	188	200	213	225	238	250	263	275	288	300	313
1:85										59	71	82	94	106	118	129	141	153	165	176	188	200	212	224	235	247	259	271	282	294
1:90										56	67	78	89	100	111	122	133	144	156	167	178	189	200	211	222	233	244	256	267	278
1:95										53	63	74	84	95	105	116	126	137	147	158	168	179	189	200	211	221	232	242	253	263
1:100										50	60	70	80	90	100	110	120	130	140	150	160	170	180	190	200	210	220	230	240	250
1:110											55	64	73	82	91	100	109	118	127	136	145	155	164	173	182	191	200	209	218	227
1:120											50	58	67	75	83	92	100	108	117	125	133	142	150	158	167	175	183	192	200	208
1:130												54	62	69	77	85	92	100	108	115	123	131	138	146	154	162	169	177	185	192
1:140												50	57	64	71	79	86	93	100	107	114	121	129	136	143	150	157	164	171	179
1:150													53	60	67	73	80	87	93	100	107	113	120	127	133	140	147	153	160	167
1:160													50	56	63	69	75	81	88	94	100	106	113	119	125	131	138	144	150	156
1:170														53	59	65	71	77	82	88	94	100	106	112	118	124	129	135	141	147
1:180														50	56	61	67	73	78	83	88	94	100	106	111	117	122	128	133	139
1:190															53	58	63	68	74	79	84	89	95	100	105	111	116	121	126	132
1:200															50	55	60	65	70	75	80	85	90	95	100	105	110	115	120	125
Radius (miles)						0.7	1.00	1.22	1.41	1.57	1.73	1.86	2.00	2.12	2.23	2.33	2.44	2.54	2.64	2.73	2.82	2.91	2.99	3.07	3.15	3.25	3.31	3.38	3.45	3.53
(km)						1.13	1.61	1.96	2.27	2.53	2.79	2.99	3.22	3.41	3.59	3.75	3.93	4.09	4.25	4.40	4.54	4.69	4.81	4.94	5.07	5.23	5.33	5.44	5.55	5.68

Figure 4-5. This table enables you to quickly determine the average herd size when considering different possible stocking rates and grazing unit sizes.

- *Topography:* Clearly, topography is an important consideration in planning the layout of paddocks and centers, even where no fencing is used. Any combination of layouts can be considered, but if you opt for a radial layout, there is an additional requirement: If the land slope is concave (most common), the radials should converge toward the bottom of the slope. If the slope is convex, then radials need to converge toward the top of the slope. In both cases, this enables the animals to move across flatter ground as they approach water (see fig. 4-6).

- *Marginal Reaction:* In general, development cost per acre or hectare goes down as grazing unit size increases—but figure this carefully. The same number of radial paddocks requires less fencing when arranged around several centers than when extended long distances from a single center. On the other hand, supplying each center with water may cost a lot.

- *Land Boundaries:* Customs and prejudices surrounding land tenure often inhibit a flexible viewpoint, but both economics and ecology frequently justify management units quite unrelated to ownership units. If circumstances warrant and cooperation seems possible, amalgamating properties for management (not ownership) may make a lot of sense. Keep an open mind.

Radial Layouts on Convex and Concave Slopes

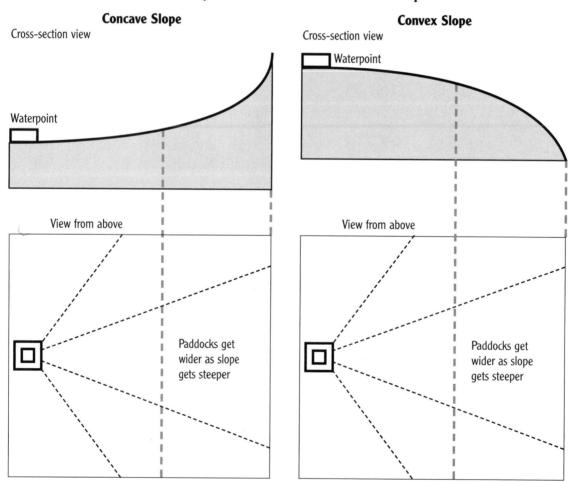

Figure 4-6. Water points should be located at the bottom of concave slopes and at the top of convex slopes in a radial grazing unit to give animals the opportunity to walk to water along the contours.

Box 4-2. Thinking Through a Grazing Unit Size/Herd Size Problem

Let's say you have land that has supported a stocking rate of 1:20 for many years under continuous grazing. Now you contemplate running herds of at least 600 head. According to figure 4-5, a 12,000-acre (or hectare) grazing unit would accommodate one herd of that size. The bottom row of the chart gives a radius—the greatest distance to water of 2.44 miles (3.93 kilometers) for a circle that big. Could your livestock handle that?

If you think so, consider what will happen when your carrying capacity increases—as you have reason to expect, judging by historical data and the performance of a neighboring ranch under Holistic Management. A fifty percent increase in carrying capacity to 1:15 will mean 800 head. Could you handle that? If capacity doubled to 1:10, could you increase to 1,200 head? And if so, how long would it take? Could you supply 18,000 gallons (68,000 liters) of water daily to 1,200 head?

If you expect to double your stock numbers, perhaps you should build two 6,000-acre (or hectare) units to take your 600 head now. What will that cost? If you build one 9,000-acre (or hectare) unit and plan the grazing, perhaps that will allow a stocking rate of 1:15 now, which would accommodate your 600 head. Then an increase to 1:10 would raise your herd to 900. Is that better than 1,200?

Let's say that water limits you to 600 head, and you see no chance of developing more. If you figure you can count on a 1:15 stocking rate nine years in ten, what options will you have when drought does strike?

This is the kind of reasoning that will lead you to a rough idea of the planning unit that best suits your conditions.

Two 6,000-acre (or hectare) grazing units
600 head @ 1:20 acres (or hectares)
30 paddocks
Distance to water: 1.73 miles
    (2.79 km)
*Fence: 74 miles @ $1,000/mile
    (119.14 km @ $621.12/km)
Two centers: $10,000
Total cost: $84,000
Cost/head: $140
Cost/acre or hectare: $7
*Assumes that grazing units don't share fence.

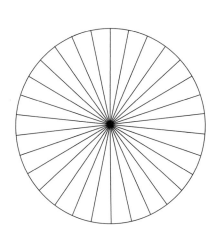

One 12,000-acre (or hectare) radial grazing unit
600 head @ 1:20 acres (or hectares)
30 paddocks
Distance to water: 2.44 miles (3.93 km)
Fence: 89 miles @ $1,000/mile
    ($89,000 total)
    (143.29 km @ $621.12/km)
One center: $5,000
Total cost: $94,000
Cost/head: $157
Cost/acre or hectare: $7.83

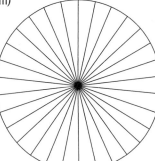

One 9,000-acre (or hectare) grazing unit
600 head @ 1:15 acres (or hectares)
30 paddocks
Distance to water: 2.12 miles (3.41 km)
Fence: 77 miles @ $1,000/mile
    (123.97 km @ $621.12/km)
One center: $5,000
Total cost: $82,000
Cost/head: $137
Cost/acre or hectare: $9.11

Creating Your Plan

WHEN YOU HAVE COMPLETED THE BACKGROUND WORK—gathered information and opinions on the status and uses of the land, recorded all your planning considerations on overlays, decided on the guidelines for grazing unit and herd size—the time has come to organize some group sessions over a day or two to generate a number of very different layouts for your infrastructure. This is a critical step, and it benefits from the presence of outsiders because, although familiarity with your land imparts a certain wisdom, it can also blind the imagination.

Generating a Series of Plans

Creative planning—especially if a number of people contribute to it—will generate far more layouts than you can afford to record on expensive maps or even transparent overlays. One solution is to get multiple photocopies of the master map you used when creating the overlays—at least ten for each planning team—and do the creative planning on them. Even copies, however, are not cheap when you need a lot of them. If the planners have access to a good map and overlays and can refer to it as they work, rough tracings of land boundaries and such major features as streams and canyons on sheets of butcher paper or newsprint will work nearly as well. Such crude outline maps are quick and simple to make—just tape a master to a large window and trace over it. If your master map does not have an opaque backing, the backlighting will allow you to trace directly from that. Otherwise you will have to make the tracing master from tracing paper or overlay material.

Finally, make one copy for each planning team of the map you created showing existing developments, and set it aside. Each team will be given an opportunity to create a final plan using this map after they have exhausted all possibilities using the master map.

Armed with a good master map, overlays, a good supply of planning circle cutouts of different sizes, cheap tracings or photocopies, pencils, and markers, you are ready to hold your planning session. Make sure you conduct this session when people are fresh and creative.

Divide the Planners into Teams

Keep the groups small—not more than three to six individuals. You might have husbands, wives, employees, tenants, neighbors, management club members, youth groups, government employees, and conservation organizations. Mix and match if you can. If you do not have many people, it is better to have four working individually than two pairs. Your goal initially, remember, is to get a large number of plans. If you bear ultimate responsibility for the final plan and you sense that others are likely to defer to you without discussion, consider not joining a team yourself. Almost certainly your ideas will turn up as elements in others' plans. Do not be concerned about excluding your most knowledgeable people, because once you have got a multitude of plans on paper, you and the people who know the land and the operation well will ultimately select the best plan, as you are the most qualified to do so.

Brief the Planning Teams

Give a general overview of the management factors you listed in the first phase of planning and how those relate to the developments being planned. Display or pass out a list of all the infrastructure needed, so the planners can refer to it throughout the session.

Explain whether you want to emphasize herding or fencing and on which parts of the land you would do one or the other. Ask for a specific number of paddocks, because high paddock numbers in particular affect the positioning of water and approaches to it within grazing units. Do this in order to force the planning teams to look very hard at the topography and complications in transportation, moving, handling, and watering stock. If you have a vision of 100 paddocks, now is the moment to figure out how to create them—even if you cannot actually put in that many for years. It may also prove helpful to show some layout possibilities to introduce the idea of handling facilities designed to serve associated paddocks in a grazing unit. (Several ideas are given later in this section.) When you do this, however, be sure to emphasize the variety of possibilities rather than any particular design.

Explain the thinking behind the herd and unit size decisions and the meaning of the grazing unit planning circles. Emphasize that at this stage you are not committed to using existing facilities, water points, or fences—though obviously if the country is notoriously dry, planners cannot put a new water point in every paddock.

Brainstorm Possible Layouts

Warm up with a brainstorming exercise. The game of listing solutions to some humorous problem, described in part 1, loosens up the mind and promotes healthy cooperation among the planners and competition between the teams.

After the warmup, provide each team with master maps (topography and fixed features only) and planning circles, and ask them to create as many possible layouts as they can for the infrastructure needed, bearing in mind all the factors mentioned. Give them plenty of time to accomplish their task.

In creating each plan, they should only be concerned with the layout of major features, such as grazing units (not necessarily all of the paddocks within them), access

to water for livestock, cropfields, working and storage facilities, landings, or staging areas for timber extraction, and so on. More detailed planning, such as the positioning of individual fences or minor roads, will be done later.

For now, you want the participants to avoid concentrating too much on any one possible layout because it may close their minds to further possibilities. As soon as any team has captured an idea of a possible layout on one of the maps, have them set it aside and start on a new one. Though this is a serious exercise in many respects, it is important that each team brings a playful attitude to it and has fun in the process. The more you can encourage this, the more creative and promising the results will be.

Create One Layout Based on Existing Facilities

When all teams have exhausted their ideas on possible layouts, give each one a copy of the map showing the existing physical structures. Ask them to creatively develop a layout for the infrastructure of the future based on the present situation. Most land managers instinctively do this first in the belief that they can derive the most "practical" plan from the current layout, because subconsciously they really do not want to spend money redoing any of it. Be assured that the existing layout, because it is based on old concepts of land and livestock management, is seldom the ideal for the long term, and it limits everyone's thinking. Thus it is imperative to generate some completely original plans from blank maps. On the other hand, the build-on-the-past option deserves consideration too, and parts of it usually wind up in the final plan.

Doing your creative planning from blank maps first in no way implies that the old layout is useless. The point is that you may want to use it differently by incorporating what you can of it into the final plan, as shown in figure 4-7. If you start from a pattern of existing fencing, you will tend to visualize that layout as grazing units and think only in terms of cross-fencing them.

Evaluate the Plans

Let the planners present their options, explaining how and why they did what they did. The point, as in the brainstorming exercise, is to find new ideas, not compete or keep score on bad suggestions. Do take notes, however,

Planning on a Sketched Map

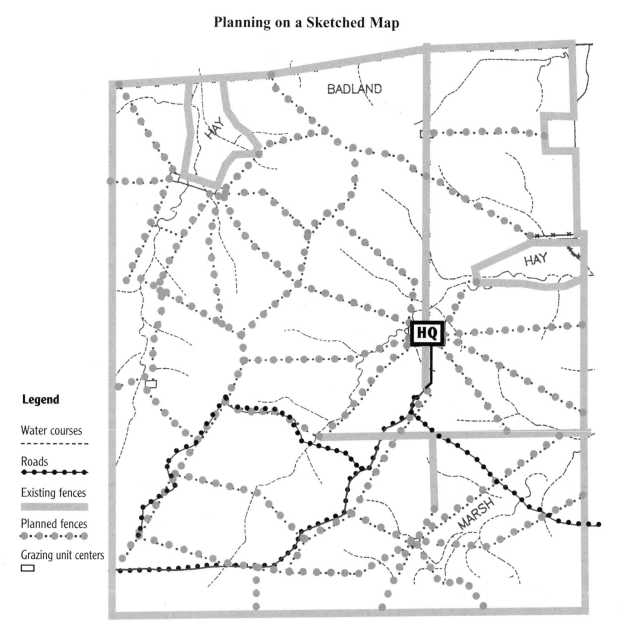

Figure 4-7. Here is a land plan laid out on a sketch traced from the topographic map shown in figure 4-2. Because the sketch shows watercourses and permanent roads, you can plan according to watershed features, even without seeing the detailed contours. Note that the plan makes use of some existing fences to define grazing units and the paddocks within them.

and use a marker or brush pen to highlight promising concepts.

After the presentations and the posting of the various plans where everybody can see them, start discussion of good elements in others' plans. The group often generates ideas at this stage that you could never have thought of on your own, so don't skip this important step. Remember,

the people with the most knowledge, and that includes you, are going to develop the eventual plan.

Designing the Ideal Plan

Transforming the best ideas into a workable plan will take some time. Those of you who know the ranch or farm well should create the ideal plan from the many possibilities already proposed. Remember, you are seeking the best long-term plan from all points of view.

A Process for Selecting the Best Features

You will probably manage to boil down your options to two or three without much difficulty by doing the following:

1. **Make map overlays incorporating your best ideas.** Superimpose them on a map showing existing facilities. Can you foresee a sequence of development that will keep prior structures in use until it makes sense to replace them, if they need replacing? Think both short and long term. For example, a capital-intensive layout may lead to higher short-term cost but much lower long-term operating costs, and vice versa.

2. **Review water supplies.** If you run your animals in a single herd all the time, or amalgamate them in a drought, do you have enough flow to sustain the higher numbers in all paddocks? If not, can you afford to haul water to deficient paddocks during the occasional year you may need it?

3. **Review your checklists.** Have you accounted for every item—all those details that were not considered when the planning teams created their layouts? For example, if many paddocks are required, certain grazing unit layouts will facilitate this while others may not. If you have planned many small crop fields that will be harvested mechanically, bulk handling from centralized crop collection points, and the routine use of livestock in the management of the crop fields, certain layouts will enhance management efficiency while others might decrease it. In either case, you are specifically looking for layouts that enable you to move livestock and/or machinery to almost any point without constantly moving over the same ground.

4. **Check each possible plan against the overlays.** As you place your original overlays—natural and social factors and the future landscape map—over each possible plan, you will begin to see both good and bad features from the point of view of each consideration on the overlay. Quite commonly at this point, you will begin to draw up new plans that combine the best features of the others. But don't rush to any final decisions yet. Let these new plans sit a while. Let others mull them over and tinker with modifications while you work through the details of cost and construction. Then come back and create the final plan.

5. **Check your ideal plan against the reality on the land.** As a final check, go out and walk the land. Compare planned fence lines and other developments with the reality of the topography, making adjustments as needed.

Implementing Your Plan

Once you have settled on the ideal plan for the future, you can begin the gradual process of changing over from the old to the new. Commonly, the cost of the changeover is a major limiting factor in the rate of change. However, if adequate money is available, the rate of change need not be slow, but it should be sound in every respect and in line with your holistic context. In creating your annual financial plan, you will be allocating money toward the desired developments in a way that differs from the conventional.

Normally you would regard most of the physical structures to be developed as capital expenditures—the capital coming from your previous earnings or more commonly from an outside source that involves repayment with compound interest. Eventually, capital expenditures yield a return on the investment, although there can be some delay before this return is seen. Provided a profit can be penciled in, the capital is usually invested. This approach can be very costly. Fortunately, on many ranches or farms, having a lot of capital available up front is no more necessary than it is in a small start-up company run out of a home. No development should cost you outside money unless you choose to apply it. Over the years, the land itself should be able to generate most, if not all, of the funds you will need. Implementing your plan in this gradual way, and generating the capital to do it from the land means that you can continue to operate at a profit year after year.

Figuring Costs and Schedules

Your annual holistic financial plans should govern the speed with which you implement your holistic land plan. If energy conversion is not the weak link in the chain of production from sun to money, then building fences is seldom wise. In many operations, product conversion will be the weak link initially—there are simply too few animals to eat the forage available. As soon as energy conversion is the financial weak link, develop the elements of your plan that give the greatest marginal reaction in that area. The science of this is not exact because of the number of variables and the difficulty of measuring them, but you can make consistently better decisions by basing them on the weak link.

Break the Land Plan Down into the Smallest Plausible Steps
Compute the cost of each step. A fencing layout might involve several stages of center construction, water development, and the fences themselves, fence by fence. There may be some items, such as a well (borehole) or pipeline, that you will have to build at once and completely in order for anything else to follow—even though you may not be able to use the full capacity of this facility for some time to come.

Such cases are costly. You will do much better if you can find a way to build incrementally so that every improvement pulls its full weight through generating more income from solar energy the minute it comes on line.

Determine a Sequence of Construction

All the advice in the financial planning section applies here, of course, particularly the three context checks: Energy/Money Source and Use, Marginal Reaction, and the Financial Weak Link. The Energy/Money Source and Use check will lead you to give priority to those elements of your plan that contribute most directly to the production of solar dollars. This could be the development of a water point that will increase your usable range—but only if you have more livestock than the present grazing area they have access to can handle.

Far more commonly, when energy conversion is the weak link, some water is available (if inconveniently so), and the highest marginal reaction comes from raising stock density and decreasing grazing periods relative to recovery periods. If you decide to do this through fencing, you will probably find it wise to do as much as you can before investing heavily in a more efficient water system. Then the question arises over which fence to build first (marginal reaction) when you have land of different productivity and an existing fence layout designed according to different criteria. The following four examples show how these factors affect practice.

1. **Fences or water?** In this example, the plan calls for converting three large paddocks into four radial-layout grazing units. The drawing in figure 4-8 shows the existing layout in bold ink and the planned four-grazing-unit arrangement in dashes. (Note that although we are referring to four grazing units, all four could be combined into one grazing unit at times.) Here are the challenges:

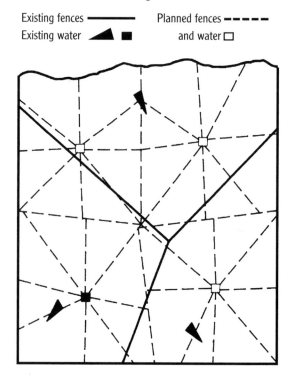

Which Facilities Would You Develop First?

Existing fences ——————   Planned fences ─ ─ ─ ─
Existing water ◢ ■   and water ☐

Figure 4-8. A number of factors should be considered before spending money on water point development or fencing, including how long you can make do with what you have.

- Assuming that developing water at each center involves fairly major capital expense, how many fences included in the new plan could you build before having to change a water point to avoid a dry paddock?

- Assuming that, aside from the water, your centers will initially contain no expensive handling facilities and consist of only a corridor and gates, which would you build first?

- In which center would you first develop the water?

- How long would you use the existing fences (dark lines) before replacing them? Doubtless you can think of circumstances that would dictate changing one of the water points earlier in the sequence, but you'd want to reason carefully.

2. **Which fence first? The Recovery Approach:** Generally, when you are constructing permanent fencing, you will decide which fence to build first according to how it will reduce overgrazing on the largest number of plants. In this example, all fences cost the same, and productivity is uniform across the planned paddock. The question is which two radial fences should you build first (see fig. 4-9 and table 4-1)?

Choice A creates two small paddocks out of the twenty you have planned. They would at once get all the benefit of higher density and shorter grazing times that someday would accrue to the whole grazing unit, though at this point there would be little impact on the rest of the grazing unit.

Choice B, also a step toward the final plan, makes four large paddocks. It will affect the whole grazing unit more evenly, but no part of it so intensely. If you work out the grazing periods according to step 11 in the Growing Season Aide Memoire (see part 2) for thirty-day and ninety-day recoveries, you will see the advantage of choice B.

Choice A gives such unbalanced grazing periods that plants will be overgrazed in ninety percent of the grazing unit—both from ani-

Which Fence First?

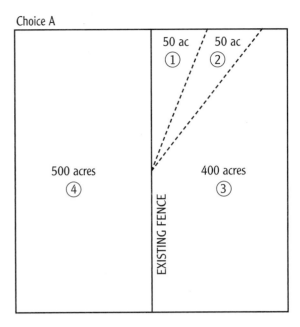

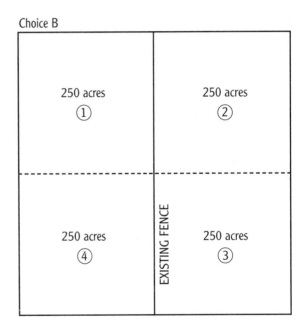

Figure 4-9. Choice A would lead to unbalanced grazing periods (and overgrazing in ninety percent of the grazing unit). Choice B would provide sufficient recovery periods, but it would not eliminate overgrazing entirely.

mals staying too long in the large paddocks and from returning too soon. During rapid growth, paddock A4 will get only twenty days' recovery time. And with only four paddocks you cannot extend that by adding days to other paddocks without stressing both cattle and land in the smaller paddocks.

Table 4-1. Grazing periods for choice B

Paddock	Min GP	Max GP
A1	2	6
A2	2	6
A3	16	48
A4	20	60
B1	10	30
B2	10	30
B3	10	30
B4	10	30

Choice B will not completely end overgrazing, because four paddocks do not shorten the grazing periods enough. Recovery periods will be sufficient, and 100 percent of the grazing unit will benefit from the recovery time. Real situations that involve land productivity, differences in fencing costs, and often many other variables are seldom so neat—as the next example shows.

3. **Which fence first? The Productivity Approach:** This example, illustrated in figure 4-10, includes the added variables of productivity and fencing cost. Common sense is often your best guide here, but the choice usually looks less murky if you can figure the productivity of new paddocks you mean to create in animal-days and divide that figure by the fencing cost. This will rate each new paddock in animal-days per dollar (AD/$)—a pretty good index of the marginal reaction of each additional fence because, normally, you want to gain the most production for the least amount of money.

4. **A composite problem:** The drawings in figure 4-11 show two possible plans for the same ranch. The broken lines show the layout of eight existing pastures being grazed as paddocks in a single grazing unit. By subdividing them you can greatly increase density and control of time. The two plans are similar. In this case we will choose plan B on the basis of

Choosing Fences by AD/$

Paddock 1	Paddock 2	Paddock 3	Paddock 4
350-acre paddocks	150-acre paddocks	300-acre paddocks	200-acre paddocks
12 ADA	25 ADA	18 ADA	12 ADA
4,200 AD	3,750 AD	5,400 AD	2,400 AD
$1,500 fence cost	$2,000 fence cost	$2,000 fence cost	$1,200 fence cost
2.8 AD/$	1.875 AD/$	2.7 AD/$	2 AD/$

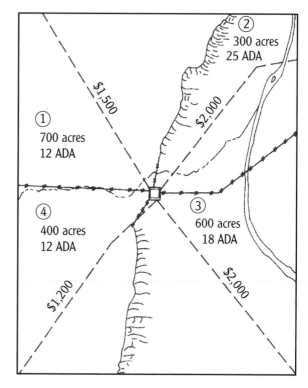

Figure 4-10. In this example you know the productivity of each existing paddock in animal-days per acre (ADA) (or ADH If you substitute hectares for acres). For the sake of simplicity, the plan calls for splitting each paddock exactly in half. Thus 700-acre (hectare) paddock 1 becomes two paddocks of 350 acres (hectares) each. (If your fence will split the paddock unequally, you must use the productivity figure of the smaller piece.)

absolute cost, though it is worth noting that the difference results from less costly water developments. The more nearly radial fence design actually requires more miles of fence. The land varies in productivity from 10 ADA(H) to 38 ADA(H), and the productivity of existing paddocks varies from 1,700 AD to 28,880

Two Possible Plans for the Same Ranch

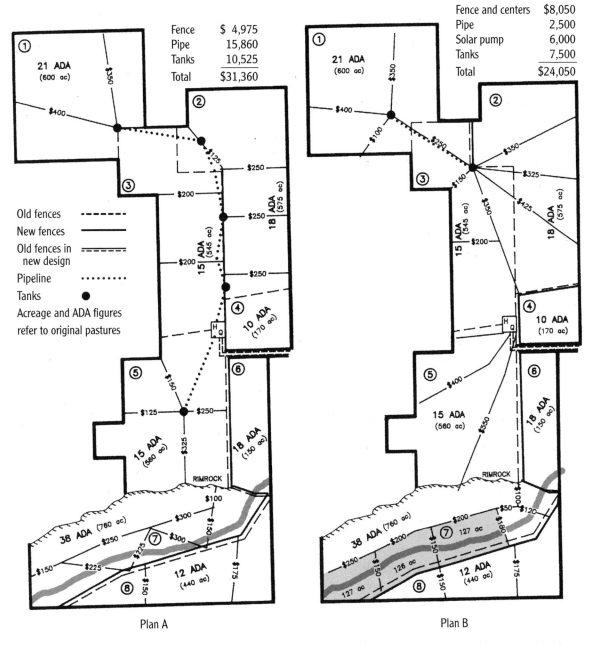

Figure 4-11. Here are two possible plans for the same ranch. The two plans are similar but plan B wins out over plan A, because water development would be far less costly. The AD/$ rule shows that fencing the shaded area first gives the best marginal reaction.

AD—as you can easily reckon by multiplying AD by acres or hectares. Having once settled on plan B, common sense suggests that you should start developing the fence layout by subdividing the highly productive riparian paddock 7.

However, by first checking your grazing and recovery periods and working it all out for a minimum recovery period of thirty days according to step 11 in the Growing Season Aide Memoire (see part 2), you would have the grazing periods shown in table 4-2 (not rounded off or adjusted) in each of the original eight paddocks.

Table 4-2. Grazing periods with thirty-day recovery

Paddock	Animal-Days (ADs)	Min Grazing Period (GP)
1	12,600	5.5
2	10,350	4.5
3	8,175	3.6
4	1,700	0.7
5	8,400	3.7
6	2,700	1.2
7	28,880	12.7
8	5,280	2.3
	78,085	

Average AD: 9,761

Average GP: 4.29 days

Clearly, paddock 7 stands to suffer from thirteen days of grazing in rapid growth, and its riparian nature will probably exacerbate the damage by assuring fast growth, even during drought. The AD/$ rule shows that fencing the shaded area gives the best marginal reaction. At 38 ADA(H)—it could be higher or lower than this paddock average—the shaded area rates as follows:

38 ADA(H) × 380 (127 + 126 + 127) = 14,440 AD.

Fence cost = $250 + $200 + $200 + $150 = $800.

14,440 ÷ $800 = 18.05 AD/$.

This would, of course, only change the grazing period in paddock 7, but there it would reduce it to 6.4 days on each half.

You can work out for yourself what the next division should be. One of the $150 fences crossing the shaded area looks like a good bet.

Allotting Time and Money for Development

All kinds of loss and headache can result from projects that come on line too late because of some bottleneck in the work itself or the financing. The procedure described in part 1 for Holistic Financial Planning shows you how to work the cash requirements of any long-term project into your financial projections. If you're careful, you should have cash on hand when you need it and know what your debt level will be at any point.

Any major construction project—such as a grazing unit center or miles or kilometers of fencing—probably deserves its own column on the financial plan. It certainly requires a separate worksheet for planning the progress of the work and no doubt a lot of extra documentation on top of that. Most project planning approaches include at least the following six steps (we provide details for working through them on paper, rather than computer software):

1. **Break down the project into separate tasks.** You can note the tasks on file cards and then shuffle them into chronological order later. It is often easier if you think backward, starting with the completed work, breaking that into large categories, and then subdividing each of those into smaller and smaller tasks. For example: A completed grazing unit center might require a water system, outside fences and gates, and handling facilities, and each of these features would be broken down into tasks and subtasks represented by separate stacks of file cards. (See box 4-3 for guidance when there is no fencing.)

2. **Assign a time requirement and a cost to each task.** For example: Order pump $1,500—delivery time three weeks; advertise for crew $200—two weeks; and so on.

3. **Arrange the tasks in chronological order.** One way to do this is to make a timeline out of adding machine tape marked in one-inch (2.5-centimeter) intervals (thirty feet, or ten meters, to the year, cut to six-foot/two-meter sections to stretch out on a table). Then lay out the cards, working backward from the completion date. This method shows graphically what tasks must occur simultaneously.

4. **Adjust the schedule.** Add slack for holidays and unexpected delays; add labor or machinery to shorten completion times where necessary.

Box 4-3. The No-Fencing Option

Although permanent fencing is not only desirable but also practical in many cases, situations exist where fencing in any form—permanent or temporary—is either impractical or counterproductive in terms of your holistic context.

For example, Dimbangombe Ranch (managed by our Savory Hub in Zimbabwe, the Africa Centre for Holistic Management) has a large and diverse game population, including elephant and buffalo, that would make short shrift of any fencing. In such cases, herding is one option to consider, and it is what the Africa Centre has opted to do.

Herding can be made much more effective when combined with the low-stress handling techniques promoted by innovators such as Bud Williams, who have demonstrated the degree of movement control that can be established with skilled handling and no (or very little) fencing. Virtual fencing, using electronic implants in the animals and movable beacons, may soon be another option.

If you do opt for no fencing, the land plan remains a vital concern. You still need to plan roads, handling facilities, and grazing distances from water, and to demarcate paddocks to control time through the grazing planning process. In the case of Dimbangombe Ranch, the land plan has relatively few paddocks—only a few are needed when you are herding—and in this case they are all demarcated by natural features.

They create their grazing plan using a slightly simplified aide memoire that minimizes the amount of arithmetic, which becomes possible with herding. (Ten or fewer large "paddocks" can be identified and a single long recovery period planned, which also results in long grazing periods in each paddock. Then herders, with their intimate knowledge of the ground, take over and graze the animals in each paddock for the planned time but do not graze any one area within it for more than about three days.) None of the sophistication of the more complex planning is lost when using this aide memoire, and the plants within any area of a paddock are subjected to a very short grazing period with high animal impact followed by prolonged recovery.

5. **Analyze the chronology for ways to improve efficiency.** Opportunities for creativity and major savings are usually enormous. They range from coordinating labor and machinery (Can the same backhoe you lease to lay pipe also be used to level the floor of the shearing shed?) to finessing transport costs (If every pickup going to town hauls back a roll of wire, can you avoid paying a trucker to fetch the whole lot?)

6. **Monitor progress.** Your chronology will of course have completion dates for all the various stages, but it is easy to underestimate construction times, particularly on long projects like fencing. Reduce all long tasks, such as plowing, fencing, or clearing, to units per day, and check the rate frequently. This will give you an early warning of delays.

Layouts and Hardware

Over the years, many holistic managers have accumulated a good deal of experience in solving the technical riddles posed by the need to control the time, density, and impact of animals. New approaches turn up continually, and occasionally old ones prove faulty as evidence matures. Recorded here are some of the more durable ideas.

Paddock Layouts

Planning requires you to mark the paddock boundaries. Fencing, of course, serves this function. For herders, a little fence can greatly speed a day's work. But rock cairns, flagging, or natural landmarks can also be used to designate paddocks.

Natural and ownership boundaries obviously play a major role in the layout of paddocks, but there is no need to organize them around different soil types. Although in the past this was often recommended to avoid damage from livestock, with Holistic Planned Grazing the time animals spend on the land is always short relative to recovery periods, greatly minimizing the chance of damage. Instead, your layout can reflect any number of social, managerial, political, ecological, or aesthetic considerations.

A grazing unit in which paddocks radiate from a common point is hard to beat for flexibility. No other design allows for such a variety of moves from one paddock to another. Nevertheless, it does not suit all situations—long, narrow canyons, for example. And under very boggy conditions, herds might concentrate too often near a center and overtrample the ground.

In principle, you want a design that allows maximum ease of movement from one paddock to any other paddock, minimum fencing distances, and most efficient use of water. Here are some thoughts and guidelines:

- Trails are most likely to form and be most damaging when livestock move up or down steep slopes in the narrow end of a paddock near the center. By siting the center on fairly level ground at the top of a convex slope or the bottom of a concave slope, as shown earlier in figure 4-6, you give animals enough space to move back and forth along contours.

- Trails are less likely to form along fences that follow the worst terrain, leaving the better country open to livestock movement. As a rule

of thumb, try to build your fences along ridge tops and straight down the points of ridges. Even so, the location of trails is often a mystery understood only by the animals. In very hilly country, you would be wise to build a few of the most obvious fences first and see how the stock move for a year or so. This trail will either confirm the rest of the plan or suggest changes. In steep country, fences constructed on the diagonal up a hill will often encourage stock to climb the hill more willingly. You might try temporary electric fencing first, as experience shows that small adjustments in location and angle often yield very different outcomes in animal behavior. Fences do not have to be straight, though bends do require straining posts.

- On highly productive land where distances are likely to be short and high density obligatory, larger paddocks enclosed by permanent fences can be further subdivided by temporary fences, as shown in figure 4-12. Because high density causes rapid depletion of forage and intense fouling, livestock will concentrate on each additional section of an area encompassed by a single moving fence almost as if a second fence were moved along behind them. The same is true in the innovative paddock layout described in box 4-4.

Paddock Subdivided with a Movable Fence

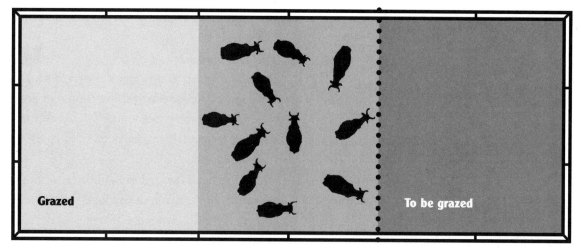

Figure 4-12. When strip grazing, it's not always necessary to have a fence to keep stock from returning to grazed areas if grazing periods are short and density is high.

Box 4-4. A Land Plan for Ultra-High-Density Grazing

In an attempt to reduce the high degree of partial rest on his ranch in Zimbabwe, Johann Zietsman found a creative way to greatly increase animal impact through land planning. He spent a little over a year creating a number of different land plans until he ended up with one that enabled him to create an unlimited number of paddocks without having to develop any additional water points, as shown below.

Johann solved the water distribution challenge by turning existing roads on the ranch into temporary corridors (see inset). For grazing planning purposes, the entire ranch was considered to be one grazing unit. Some of the "paddocks" (A–H) were demarcated by permanent fence (which was being taken down); others were merely imaginary lines. Each paddock was subdivided into strips (1, 2, 3), then each strip divided numerous times into temporary mini-paddocks—up to 1,500 in all, giving him tremendous flexibility. When he hit a drought in the

first year he had more than enough paddocks to ensure animals could move onto fresh forage throughout the day every day, without having to return to a paddock until the rains finally came many months later.

In the inset, strip 1 has been grazed; the main herd—545 cows, calves, and bulls grazing at a density of 1,000 to 3,000 animal units per hectare (400 to 1,200 per acre)—is moving through strip 2; 91 yearling heifers he's planning to breed preceded the main herd in strip 2 and have now moved on to strip 3.

The grazing chart only lists paddocks A through H, since the chart isn't set up to plan or record 1,500 paddocks and moves that occur every few hours. Many would argue that grazing planning becomes unnecessary when you're dealing with this many paddocks. That would be the case if all you were concerned about was overgrazing, but far more is involved in creating a landscape and a quality life.

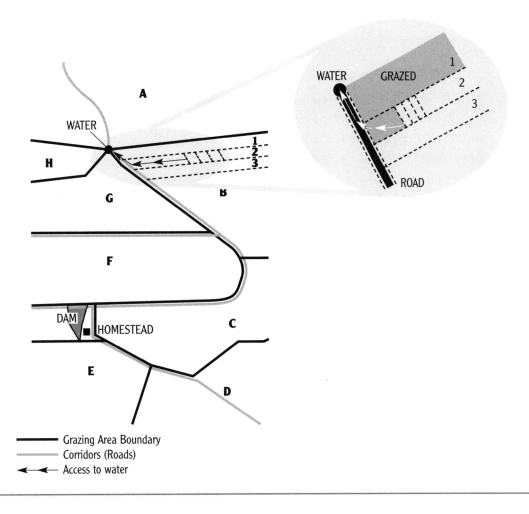

Grazing Unit Centers

In most situations, you should build centers before developing paddocks. If you put up paddock fences that have no function until connected to a center, livestock will never respect them. In any case, never introduce livestock to electric fences that do not carry full current.

In well-watered, highly productive pastures where distances are short, initial paddock numbers necessarily high, and existing facilities adequate, you may choose to delay building a new center. Nevertheless, if your ultimate plan calls for one, design it thoroughly right at the outset, and make sure that it will fit gracefully into your construction sequence. Few rules govern center design, though long experience has uncovered a few ideas that will help you avoid the most common problems (see fig. 4-13).

The first grazing units for cattle had many problems. The cattle came in to drink and stayed instead of going out on the range and getting fat. Also, the center filled up with manure that smelled bad and caused disease. To solve this problem, ranchers started building centers like the ones shown in figure 4-14.

Although one center is for cattle and the other for sheep, both have a narrow "corridor" around the outside. It should be wide enough for a pickup to drive through, but no more. Stock will not hang around too long in a narrow corridor, and will leave their manure out in the paddocks, where it will help the land, but there is plenty of room in the middle of the center for all kinds of handling facilities. Most ranchers build the corridor first and little by little add facilities inside. (See box 4-5 for some good ideas on handling facilities). When you build, always look ahead and leave space for all future needs.

Positioning Centers

When centers enjoy broad, level approaches, less trailing results—though if you plan one around an existing water point where trailing was not a problem under continuous grazing, it should not become one now. Planned grazing will improve the situation no matter how the fences run. Your next considerations must be access, water availability, construction obstacles, and proximity to roads or hazards. Many old corrals were built in potentially boggy areas because of their access to water, but you may spare yourself decades of grief if you can find a way to build on higher, drier ground. You may want to build near a well-used road—but how near? Will every passing truck startle your animals?

Center Corridors

Ideally you want cattle out in the paddocks, not lounging around piling up dung and fly bait in the center. Yet you also want space in the center for handling animals, storing supplies, and corralling riding horses, bulls, and sick stock. A narrow corridor built around a central area will solve both problems. The corridor communicates with all paddocks served by the center and can contain water facilities. Animals that come there to drink will have no incentive to loiter in the confined space and chew cud, but will move back out on the range where their dung is an asset. Meanwhile, you can enjoy a large clean central working space. The corridor might even surround an entire homestead.

Generally, you will figure the size of the central space according to the facilities you want to put there, but it is easier to be generous at the outset than to expand later. Five thousand square yards (or meters) will easily accommodate the needs of a herd of 500 to 1,000 cattle that need daily handling.

Plan the center for all the paddocks, herds, and facilities you might ever want to handle—even if you initially build only the corridor around a large empty space. The corridor should be wide enough to drive through, but fifteen feet (five meters) will accommodate herds of almost any size, and the benefits disappear entirely at about thirty feet (ten meters). The perimeter should be long enough to accommodate plenty of gates. If you are serious about production from the land you can expect to have thirty or more paddocks at some point.

When you have planned your center on paper, lay it out on the ground with pegs and twine or powdered lime and test it as far as possible. If vehicles will have to pass through it or turn around to back up to a loading ramp, be sure they can do it before you start digging the postholes. Remember that the part of the center facing the portion

Grazing Unit Centers and Fence Patterns

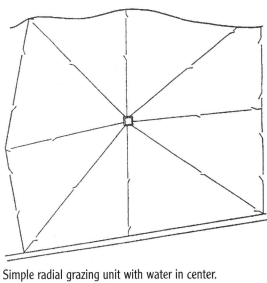

Simple radial grazing unit with water in center.
Note gates away from center

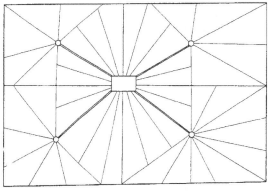

Dairy unit in center and water at satellites.

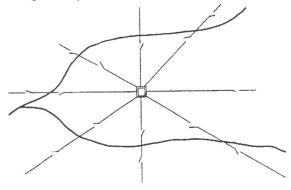

Dry center—water available in all paddocks.

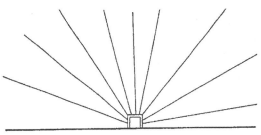

Split center:
Full facilities would be in the half serving the largest area.
Half centers can be set well back from river or road.
1-, 2-, 3-, or 4-way splits are possible.

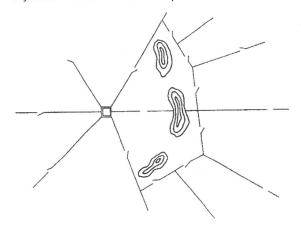

Fencing to avoid problem piece of land near center.

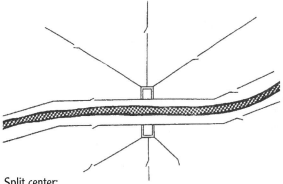

Radial fences in a fan design (takes more
fencing than wheel layout).

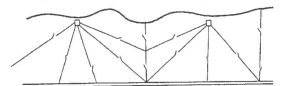

Several fans used in a narrow area.

Figure 4-13. Careful thought should be given to positioning grazing unit centers and the fences that lead into them. The examples shown here show a number of possibilities that vary depending on the terrain, the operation, and the cost.

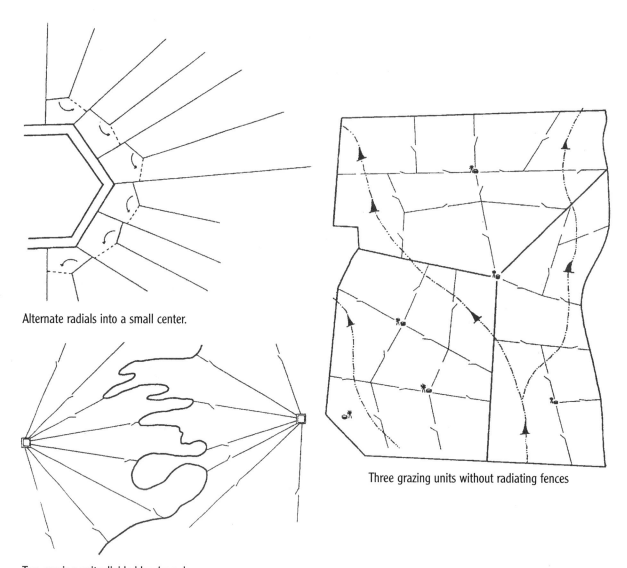

Alternate radials into a small center.

Two grazing units divided by rimrock.

Three grazing units without radiating fences

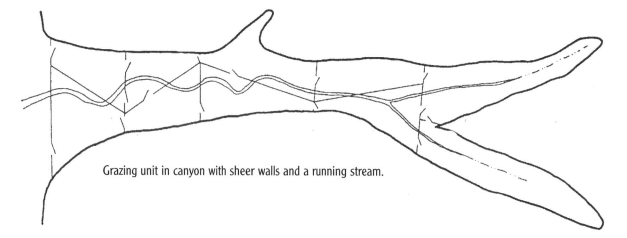

Grazing unit in canyon with sheer walls and a running stream.

Figure 4-13. *Continued*

Simple Grazing Unit Centers

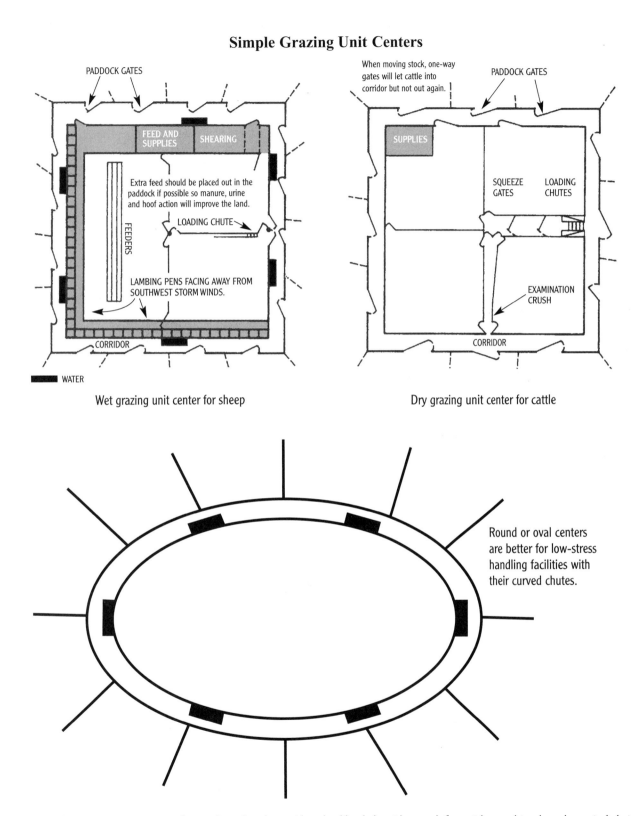

Wet grazing unit center for sheep

Dry grazing unit center for cattle

Round or oval centers are better for low-stress handling facilities with their curved chutes.

Figure 4-14. Grazing unit centers can be any shape, but the corridors should only be wide enough for a pickup to drive through, particularly in "wet" centers that serve as water points. Stock tend to loiter in wide corridors, leaving dung where it becomes a liability rather than the asset it is when out in the paddocks.

Box 4-5. Handling Facilities—Good Ideas

Consider animal stress in designing your facilities. Animal behaviorist Temple Grandin offers free designs on her website: www.templegrandin.com and in numerous publications available for purchase. Her curved chute layouts reduce stress and enable you to work animals far more efficiently.

Over the years, ranchers and farmers have developed a variety of innovations that have greatly increased management and handling efficiency. Consider the following.

Lambing Boards

Newborn lambs and kids can remain at the grazing unit center until large enough to discourage predators. To keep them there, merely place a board across the entrance to the paddock in use that is high enough to prevent young ones from jumping out until they've reached a certain size, but low enough to let mothers in to suckle.

Calf Traps

If you need to frequently put mother cows through a dip or spray race and want to separate the calves quickly and with minimal stress, here's an idea developed years ago by Zimbabwean rancher Bob Rutherford, which enabled one man to bring in 500 cows, separate their calves, and

have the first cows through the plunge dip and grazing in their new paddock in three to five minutes. And there was no bawling, or any attempt by cows to look for their calves—as long as the whole herd was back in the paddock within an hour. After a brief training period, cows got used to their calves leaving them as they approached the dip entrance and then joining them in the paddock later.

The grazing unit center diagram shows how the trap works. The cows with calves are moved through the outer corridor to the gate at point A. As the cows are compressed toward point B, the calves automatically dart back and duck between the poles of the trap (D). These poles are wide enough apart that a calf, but not a cow, can pass through. Once some calves see others in the pen they tend to move to it. After a few handlings it becomes routine, and calves simply duck into the calf trap and cows go on toward C and the dip, spray race, or whatever.

The calves in the trap are then driven through the corridor and out at point E to join their treated mothers. When calves require treatment—dehorning, castrating, and the like—this is done within the corridor to E.

Similar calf traps have subsequently been designed—some in round or oval centers—but always

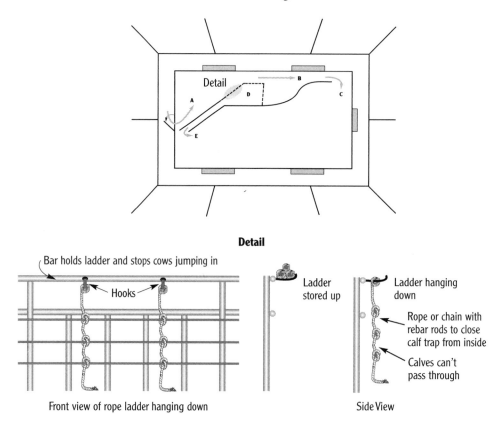

Detail

Front view of rope ladder hanging down

Side View

Box 4-5. *Continued*

with the same principle: a crowding area and the chance for calves to escape backward or sideways through the bars and into the trap.

Once in the trap, calves can be contained by simply lowering a ladder, made of nylon rope or chain and rebar rods, *on the inside* that blocks the entrance spaces between the upright poles (see detail). Because some cows do get agitated in the early training period, it's a good idea to add one high cross bar (from which the ladder descends) to prevent a cow jumping into the trap.

Single-Sire Breeding

The desire for single-sire breeding is often quoted as a reason why many small cow herds are essential. But since the resulting loss in forage production is so high, consider this alternative, developed by Allan Savory some years ago when working with a registered (pedigree) Simmental breeder in Namibia. Not only was this ranch able to run all 500 breeding cows in a single herd, but they also found they could use one bull to ninety cows (rather than the usual practice of one to twenty-five cows). What's more, they also knew the exact date on which each cow conceived and also knew quickly if any bull was faulty, without any need for bull testing.

Here's the layout used at the grazing unit center.

The bulls were placed in separate breeding pens with color-coded gates, as some of the herders were barely literate. Each bull had a colored ear tag in each ear (because the bush was very thorny and dense) to match his pen color. Although the bulls were fed in their pens, this color-coding enabled the bulls to be taken out together each day to water and to have a good walk in a paddock well away from the cows.

Early each morning, the herders simply watched for any cows in season that came into the center close to the bulls. Each cow, like the bulls, had two colored ear tags. If they were red she was put in the red bull pen, and so on. After she was served the herders let her out and noted her number and the date in a book.

The rancher was able to use one bull to ninety cows because a bull did not serve the same cow repeatedly. The date of conception was known if that cow did not return to the bull. And if the bull was defective it showed in repeat visits by the cows assigned to him.

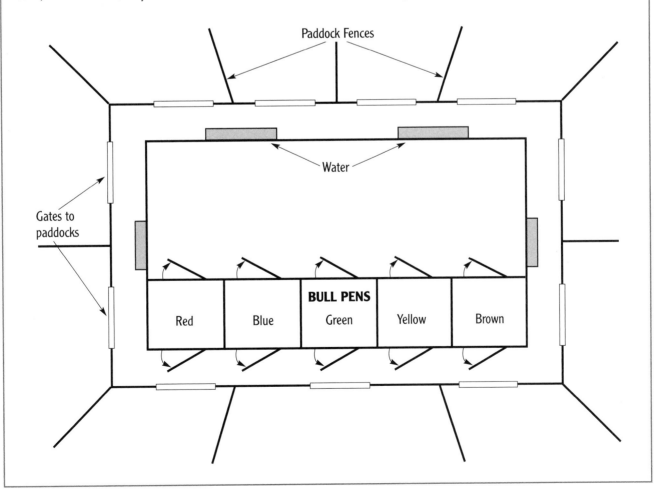

of the grazing unit with the greatest grazing area will tend to have the most fences coming in to gates around the perimeter. Lay out radial fence lines from the actual gates rather than from a common peg in the center—otherwise they will all be bunched together.

Water

If you have a winding stream or a dam that supplies all your paddocks and you plan to use them as your water point, bear in mind the following:

- Flowing river water is ideal if you have it, but it necessitates construction of walkways to prevent livestock overtrampling the bank and muddying the water.

- Dams, sloughs, and other catchments present a special challenge—especially if you use them to supply several paddocks so that stock have access to them over longer periods. Consider fencing them so that animals cannot stand in them or wallow and foul them, as shown in figure 4-15. You can drive several permanent posts that will allow you to change the access area occasionally or extend it when the level drops. Or consider siphoning water to a trough if the terrain allows.

Fencing to Protect a Dam or Small Pond

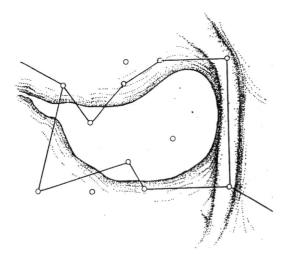

Figure 4-15. Moving wire to allow access to water at different points allows management of shore communities while preventing animals from standing in the dam or pond and fouling it.

If you do not have a stream or dam available, and your water source is a well (or borehole), you will probably want to have water piped into the center. There are more combinations of solar, wind, and gas-powered pumps, rams, siphons, and pipes to help deliver water than can be described here, and the technology keeps improving. Even if you do not have water where you want to put a center, do not abandon the idea without an up-to-date assessment. Keep these principles in mind:

- You must have a flow such that livestock never have to wait. Otherwise you will have animals packed into the corridors sniffing at drips. The dominant individuals will develop a persistent habit of driving others away. If all your livestock can trust the supply, they will only come to the center when genuinely thirsty and then leave. This means you will have to engineer a sufficient direct flow or make sure that storage capacity is big enough to handle peak demand. Where water sources are poor, it can pay to plan grazing units and then pipe small amounts of water from several minor sources to a reservoir at or near the grazing unit center. This way you can supply a herd big enough to do the land some good. (See box 4-6 for an inexpensive reservoir idea.)

- Large herds will require large amounts of water delivered, and more so in hot weather (at least fifteen gallons, or sixty liters per animal per day for cattle); do the arithmetic with herd size in mind.

- Rate of water delivery to a trough is more important than having large troughs, even with large herds.

- Depending on water source and pumping, it is generally wise to have several days' water stored in case of breakdowns; the farther away you are from parts suppliers, the more days of storage you will need.

- Livestock will perform better with cool, shaded water than with hot water. Consider adding shade over large expanses of water whenever possible.

Box 4-6. A Low-Cost, Low-Maintenance Reservoir

Eggshell-thin cement reservoirs resist cracking and can be built large enough to meet the storage capacity requirements on almost any ranch or farm. Though, "eggshell-thin" is an exaggeration, the secret is in the thickness of the cement walls—which should be no greater than 1.5 to 2 inches (35 to 50 millimeters). In the early days, the recommendation was to build the walls no higher than 6 feet (1.8 meters). However, the Africa Centre for Holistic Management built its main reservoir walls 10 feet (3 meters) high, and the reservoir has served them well for many years. The construction principles are similar to those used in building concrete yachts. Water troughs can be built eggshell-thin too.

Reservoir construction is easy: lay a foundation floor and erect the sides with weld mesh wire of suitable size. Wrap chicken wire netting around the structure and then plaster it, making a thin eggshell concrete structure.

Materials Needed

- Rolls of weld mesh wire of suitable size to form the stand-up structural base of the wall
- A role of fine (chicken) wire netting equal to the reservoir's circumference.
- Binding wire to attach the netting to the weld mesh form
- Cement, sand, water, and helpers
- Plastering trowels and a straightedge (if you want to work the wall to an even finish)

Instructions

1. Lay a circular foundation approximately 4 inches (100 millimeters) high by 1 foot (300 mm) wide at the top, which needs to be level. (If you're building on sandy soil, cover the inside floor of the reservoir with rocks and reinforcing. Then pour a weak sand/cement mixture, at 10:1, over the rocks to almost cover them—this will reduce the amount of concrete you need in the next step.

2. Set an outlet pipe into the foundation. Then concrete in the reservoir floor—3 to 4 inches (75–100 millimeters) thick. (Remember that cement should not dry but rather be kept wet until it "cures" or is well set—the longer it remains wet the stronger and less likely it is to crack.)

3. Assemble the weld mesh foundation—fortunately a weld mesh roll easily forms a circle and stands on its own. But, you may want to anchor the mold with steel stakes in case of wind.

4. Wrap the wire netting around the weld mesh and attach it firmly with binding wire.

5. Plaster the outside, using a 1:6 cement/sand mix so that it covers the netting but is not too thick. You must do this in one session, or cracks will form, so make sure you have enough cement, sand, water, and helpers on hand before you start plastering. If you want a smooth finish, use a straightedge to smooth out the bumps. Keep applying a light spraying of water to prevent the cement drying for as long as possible. Let the plaster cure wet overnight.

6. Plaster the inside using a 1:4 cement/sand mixture, again in one session, keeping it wet as you go. Then immediately begin filling with water to half full or higher to keep the cement wet and curing for up to three weeks to create a wall of great strength that is unlikely to crack.

7. Optional step to harden the surfaces: Prepare a mixture of 5 gallons (20 liters) of water, 2 tablespoons of caustic soda, and 2 pounds (1 kilogram) of salt. Let this stand overnight and then add cement until the mixture is thick and creamy. Paint the inside and outside of the reservoir with this mixture. Then keep both surfaces damp while the cement cures.

8. Fill the reservoir and enjoy years of trouble-free use.

- If you are piping water and want to use portable troughs to help spread animal impact, consider using T-junctions rather than joiners when you lay your pipe. If you place the troughs at high points you can more easily bleed any air trapped in the line. As with any other water points you will need to ensure an adequate delivery rate. Some managers have used "tapping saddles" that clamp onto a pipeline to create a T-junction anywhere they want a take-off point; some tapping saddles are self-sealing, with a spring-loaded valve, which eliminates the need for a tap. Given more take-off points you have the ability to move the trough within a paddock not only to spread animal impact but also to better utilize forage or reduce trailing.

- In some circumstances you might consider using water trucks with a trough on the back as a way to provide water, but also to move livestock to distant areas.

- The concentration of animals at the center can create enough dust to affect large open troughs. Long narrow troughs, served by a rapid flow, do better and are easier to clean, especially if the ends slope. While it doesn't seem to be a problem with lambs and kids, calves may get pushed over and drown in troughs over nine inches (twenty-three centimeters) wide, so bars should be welded across the top.

- Consider building ramps or steps on reservoirs and troughs (inside and outside) that allow birds and other small animals to drink without falling in and drowning. Not only will you keep the water unpolluted, you will be helping to build up complex communities.

- For the many shy creatures that cannot or will not brave grazing unit center fences to get water, seeps or pools are easy to construct. Figure 4-16 shows one made with a regular float valve set in half a steel drum. This device functions like the tank on a toilet. The flow runs from the covered tank into the other half of the drum buried to ground level and filled with gravel and sand. Wild things can always scrape a bit to find a drink. An old trough or plastic-lined depression would provide an even better seep than half a drum.

FENCES

Fencing technology is evolving so treat the points here as lessons learned and keep up to date with new ideas through the Savory Institute and the many publications, such as the *Stockman Grass Farmer*, that devote articles to the subject.

For some years now the solution most cost efficient, versatile, and least damaging to wildlife has been the free-running slick wire fences developed in Africa and electrified by high-voltage pulses from New Zealand-type

Water Seep for Wildlife

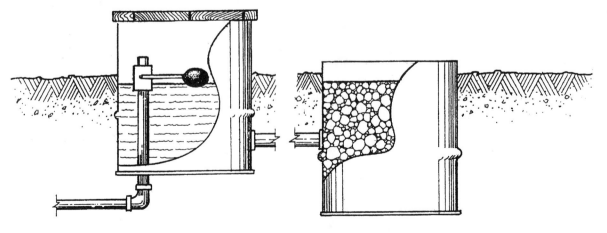

Figure 4-16. The water level in the left barrel will maintain the level in the sand and gravel on the right.

energizers. There are numerous suppliers of electric fencing materials as well as methods for powering energizers—powerline electricity, periodically replaced car batteries, or solar-charged batteries.

Posts used in electric fencing do not need to be as sturdy as those used in conventional fencing because it is not the strength of the wire or tape that holds animals back, but the fear of a shock. Should a fence prove to have been put in the wrong place, which sometimes happens, particularly in hilly country, it is not a large or costly task to realign an electric fence. It is generally not a good idea to use electric fencing where animals are forced to touch it, as in grazing unit centers where crowding animals often get pushed.

How Many Wires?

Most people start out using far more wire than necessary, much to the delight of the fencing industry. On dry soil where grounding is poor, two wires, one hot and one grounded, will contain almost any domestic stock besides goats—if they are not seriously stressed and if they are well trained to respect the fence. Where grounding is good, one wire will suffice.

Frequently you can tell the sequence of fence building on a ranch by counting the strands of wire. The early fences have three, four, or more strands and closely-set posts. Eventually confidence grows to the point where the most recent fences have two strands and posts set so far apart that the wires droop between them.

Spacing the Wires

No ideal spacing has yet been determined. Most reasonable ideas work, so the question is not critical. A common arrangement for cattle is two wires three inches (seven and a half centimeters) apart and between twenty-four and thirty inches (sixty and seventy-five centimeters) above the ground. The top wire is hot and the bottom one grounded. For smaller stock the bottom wire should be lower. You should consider three wires in the case of mixed herds of cattle and sheep. With sheep or goats in the herd you might want to make the bottom wire the hot wire to discourage animals from creeping under the fence. Contrary to conventional wisdom, the same fencing that divides paddocks will work on the grazing unit perimeter. Except for paddocks actually containing livestock, power

will be switched off in most fences, including most of the perimeter fences, most of the time.

Posts

Almost any post can be used—wood, steel, or fiberglass. Usually cost and availability will make this decision for you. Poor-quality fiberglass posts will split out, decay in sunlight, or be sawn apart by the wire vibrating in the wind, so ask for references. Most fiberglass posts and the lighter wooden posts will slide up and down in their holes. This, however, seldom causes any problem. There are many cases of deer learning to lift up fences and pass under. The posts simply drop back into place. You can also use growing trees. If the wire bends around the tree, you can often run it through plastic insulator hose. The tension will hold it in place without any need to risk the tree at all.

Attaching the Wires

The wires must run free, and the attachment must be strong, durable, and easy to install (see fig. 4-17). New devices turn up every day. The wires must be able to slide unrestricted. Otherwise any shock or expansion and contraction of the wire will concentrate its force on a single post. A free-running wire distributes the force through the whole fence to end strainers designed to take it.

Attaching Free-Running Wires

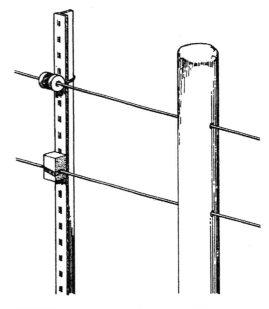

Figure 4-17. Wire attachments must be strong and free running.

If you are using temporary fencing (or permanent posts and temporary strands) that involve frequent moving of the electric wire or tape, breaks will develop, and in the rejoining, knots will appear here and there. This can lead to headaches when winding up the tape or wire as it will catch on the eyes of the posts. To minimize the problem, use eyes that allow you to lift out the tape or wire and drop it to the ground before you start winding up.

Beware of any connectors, or eyes, of weak design, especially plastic ones that become brittle in sunlight. Fiberglass and treated wood posts threaded on the wire work well, but you will have to cut the wire to replace damaged posts, and such arrangements are only suited to permanent electric fence where the wire is not repeatedly wound up.

Straining Posts

At each end of the fence and where there are significant bends, you will need a straining post. Along a long length of fence, you may want some substantial posts here and there, but most can be lighter, and they are only there to hold the wire up (as opposed to taking the strain of a high-tension fence).

The design shown in figure 4-18 is a bit simpler and quicker to build than the conventional H-strainer used in barbed-wire fences. Most experts recommend a diagonal of eight to ten feet (two to two and a half meters). Set the post deep. There should be as much below ground as above. The pad at the end of the diagonal can be rock or treated wood. Generally, a single stout wooden post can handle strain in the middle of a fence because the pull on opposite sides cancels it out. Two electric wires only add up to about 400 pounds (180 kilograms) anyway.

Grounding

One of the most common faults with electric fencing is poor grounding, especially in dry climates. If your soil and air are dry, you may well need three or four six-foot (two-meter) steel rods driven into the ground and attached in series, as shown in figure 4-19. These are connected to the charger and the ground wire in all fences. When the charger is located at the center, the power is often distributed to radial fences through wire strung around on posts above the paddock gates. If the ground wire is attached above the hot wire, it also serves as a lightning deflector.

Straining Post

Figure 4-18. This strong, simple straining post is quicker to build than the conventional H-strainer used in barbed wire fences and is all you need.

Gates

Although stock move a lot, most gates will only be opened and closed about a dozen times a year unless they are on well-used roads. But for the sake of flexibility you cannot have enough gates. You absolutely must be able to move livestock from one paddock to the next at several points other than the center. Otherwise you will train them to a destructive routine or sacrifice a piece of ground. That said, any kind of gate will work. The two-strand fence does not require anything elaborate.

Wherever your ground is level enough to allow a good distance between poles, you can push the wire to the ground with your foot. A couple of hooks buried in the ground will hold it down and horses, trucks, or herds can pass over. (See box 4-7 for another idea.)

Gully and Stream Crossings

The electric fence handles these easily. The wires pass straight over and have alternating hot and cold wires or fine chains dangling down into the gap. This barrier prevents stock from moving past but allows water and debris to wash through without damage.

To keep the drop wires from shorting, run the fence wires side by side on opposite sides of wooden

Wiring a Grazing Unit Center

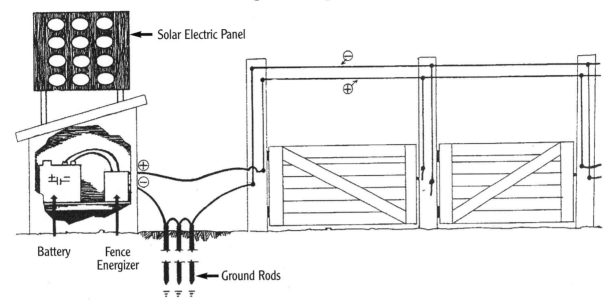

Figure 4-19. General wiring pattern for fences radiating from a center.

posts instead of one above the other as they cross the gap. Wires or chains hung every foot or so on opposite wires will make an effective barrier, as shown in figure 4-20.

Educating Livestock

Animals usually learn quickly to respect electric fence. Occasionally, however, it may help to tempt a particularly unruly lot of goats, for example, to lick aluminum cans smeared with molasses and clamped to the hot wire. In dry climates where grounding is poor, this method works best on a wet day or where the cans hang so the animals are likely to touch the ground wire in reaching for them.

Many people who annually acquire great numbers of untrained stock like to start them out in a small training enclosure at enough density to make sure that a good number of animals get zapped. Try adding an extra wire or two in the fence and hosing down the ground to increase the likelihood of a powerful shock. In any case, retaining a few experienced lead animals each year greatly speeds the education of the rest.

Training stock to move to a signal such as a whistle, bell gong, or horn also adds immensely to the ease of handling animals. Not only does it make your fencing more

effective and secure, it is also useful for concentrating livestock to produce herd effect.

Training takes a little patience and a little art. Here are the ground rules:

- Choose a unique signal that the animals only hear when you mean to move them.

- Always associate the sound with a reward, such as hay or supplement cubes. Eventually animals will find sufficient reward in a new paddock, but make the message absolutely clear while you are training them.

- Do not mix reward and punishment. If you drive them while blowing a whistle—even if you drive them toward a reward—they will not get the point.

- Consider taking a training course in low-stress animal handling from innovators, such as those who have been trained by Bud Williams, and you will never need to "drive" animals anywhere.

- Mix older, trained animals with newcomers until the message gets across.

- Start the training in smaller paddocks where the animals can see you and each other.

Box 4-7. Creating a Temporary Gate—Anywhere

If you're working with permanent one-wire fences and you want to create a gate to give animals access to water, or for any other reason, all you need to do is prop up the single wire about six to seven feet (two to two and a half meters) off the ground with an insulated pipe (PVC works well) wherever you want to make an "opening." I first saw this in operation in the humid Pampas of Argentina where they were finishing cattle. They don't mess around with making any gates at all.

Materials Needed

- PVC pipe, 6-7 feet (2–2.5 meters) long; 2–3 inches (50–75 millimeters) in diameter
- One bolt, 8 inches (200 millimeters) long; ½–¾ inches (13–19 millimeters) in diameter
- High-tensile wire (a little more than double the length of the pipe)

Instructions

It's necessary to electrify the pipe so the animals don't rub against it and knock it over. Here's how you do it:

- Drill a hole through the pipe about halfway up its length and insert the bolt through it.
- Drill a hole through the bolt at each protruding end, large enough for a high-tensile wire to pass through.
- Drill a small hole through the bottom end of the pipe.
- Cut a groove across the top end of the pipe.
- Pass the wire through the hole in the bottom of the pipe, through one end of the bolt, then through the groove on the top, then back down through the other end of the bolt, and tie in to the other end of the wire near the bottom of the pipe. When you prop up the fence wire, it will contact the pipe wire in the groove, and be electrified.

Cattle that are well trained to an electric fence might not walk through the "gate" immediately, even though it's six feet (two meters) or more off the ground, but a few brave ones will figure it out within a few minutes, and the rest will follow.

We sure would have built a lot fewer gates on our place if we'd known we could make a gate anywhere we want just by lifting the wire.

— Jim Howell, Colorado, USA

Electric Fence Crossing a Stream

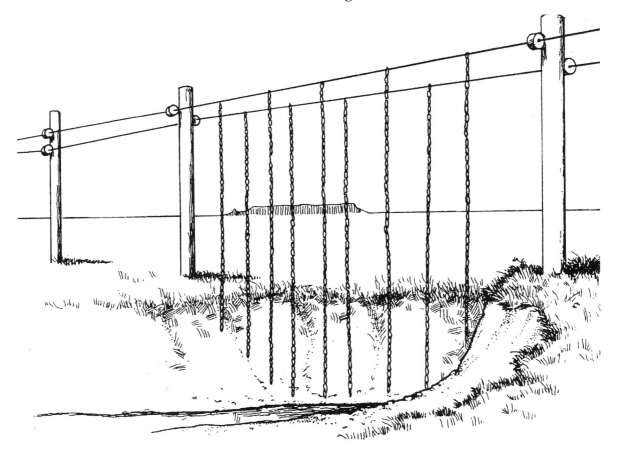

Figure 4-20. Chains hanging from parallel strands overhead create a hot barrier.

Summary

In several places we've used the metaphor of the artist standing before a blank sheet in order to inject a sense of drama and romance into tasks that will probably keep your hand on the calculator and your nose in the paper longer than you ever thought you could bear. Certainly any good management, not just Holistic Management, requires a lot of organizing, planning, and monitoring. But painters and poets don't really enjoy much drama and romance in their day-to-day work. Most grind out their lives in risky, low-status toil, and they die broke far more often than ranchers and farmers do.

Nevertheless, the comparison holds in other ways. In the blood of all who take responsibility for blank pages or pieces of land runs the desire to bring forth from them something altogether new and of lasting value.

Who knows why? A hundred times you've heard faded men in faded denim say, "I don't know why I stay in this business"—though you know that leaving the bone-grinding labor on their land would kill them in no time at all.

Because you do know why, you'll compose a land plan worthy of your art that will sustain you and others who live from it. You will create it from a fresh map, unencumbered by the roads, fences, ditches, and other clutter you inherited or built without forethought. Your land will become what it ought to become.

We've already hinted at the material benefits of planning. Too many people forfeit these benefits by clinging without reflection to what always was. A little bit of patient planning may return hundreds of thousands of dollars over the years in savings and increased income from fewer roads, water points, and fences, better livestock handling, greater wildlife productivity, reduced losses from weather and predation, and a host of other factors.

There is a bigger reason, however.

While theologians debate the immortality of the soul, you can walk around and get your boots muddy in the ages of the earth, which compared to your short life might as well be immortal.

Then again, you aren't sure, are you?

No other living planet has yet been found in space, and ours is very small. We tax it sorely with our bombs, wars, fumes, and fires, our cutting down and building up, our teeming cities and plundered fields, our grasping and our greed.

And that is why you will sit down with planning sheets, your computers, and your maps, and do your work—so that when the paintings in their galleries and the poems on their shelves have gone to dust, the earth, your piece of land, will abide.

APPENDIX 1
THE CONTEXT CHECKS IN SUMMARY

There are seven checks used in decision making, which the text, *Holistic Management: A Commonsense Revolution to Restore Our Environment*, describes in detail. Each check incorporates one or two questions you ask yourself prior to implementing a decision to ensure the decision is economically, environmentally, and socially sound, relative to your holistic context.

Cause and Effect

Does this action address the root cause of the problem?

The cause and effect test is one that carries considerable weight when a decision is taken to address a problem. It enables you to winnow out actions that only suppress symptoms when you need to correct the cause.

This check helps you avoid a recurrence of the problem by first asking you to think carefully about what might be causing it. If the action under consideration addresses that cause then it passes this check. All too often we tend to address the symptoms spawned by a deeper underlying cause because the problem (symptom) is always visible. When cockroaches appear in the kitchen, we spray them first and then only sometimes ask what has caused them to show up in the first place. Quick fixes, such as spraying a pesticide, that tackle symptoms are very attractive, because they often remove the symptom right away, but they can also produce unintended consequences. In the case of the roach spraying, the poison used can harm roach predators that help keep roach numbers in check, as well as the people using the kitchen. And if the cause of the roaches' presence isn't addressed (keeping surfaces clean, sealing off entry points, etc.), the roaches (and the problem) return.

Weak Link

A chain stretched to breaking will, by definition, fail at the weakest link. At any moment in time every chain has one, and only one, weakest link that alone accounts for the strength of the entire chain, regardless how strong other links might be. To strengthen a chain when resources are limited, one must always attend first to the weakest link.

Other links, no matter how frail they appear, are non-problems until the weakest link is fixed

The undetected weak link can cause mighty undertakings to suffer continual setbacks or even to fail outright. Thus the weak link check helps ensure that your actions address the link that is weakest at any moment. The check applies in three different situations: social, biological, and financial.

Social Weak Link

Could this action, due to prevailing attitudes or beliefs, create a weak link between us and those whose support we need?

You, and the future you envision for the whole under your management, are linked by a chain made up of all the actions you will take to get there. Any action that runs counter to prevailing attitudes and beliefs is likely to meet with resistance, creating a blockage that, if not addressed, will at some point become a weak link between you and the people whose support for your efforts is important, even vital.

If such a potential weak link is identified then it's important that you address it *before you take action.* Thus this question is not asking simply for a yes or no response but reminding you to address a potential social weak link before it becomes one. You want to ensure that as you move forward you do not inadvertently create disagreements that could have been avoided, or conflict or resistance that later hinders your progress.

Biological Weak Link

Does this action address the weakest link in the life cycle of this organism?

The weak link check applies when you are dealing with populations of plant or animal organisms that have become a problem, either because they are too many or too few in number. Before you take any action to increase or decrease their numbers, you need first to ensure that it addresses the weakest link in the organism's life cycle. In doing so, you are likely to maximize the effectiveness of the action and to ensure the results will be lasting.

Every organism has a point of greatest vulnerability during its life cycle—a weakest link. Recognize this and you increase your ability to effectively decrease or increase that species' ability to recruit new members to its population. When action you are contemplating addresses that weak link, it passes this check.

Financial Weak Link

Does this action strengthen the weakest link in the chain of production?

Each year, in conjunction with Holistic Financial Planning, you need to identify the weak link in the chain of production that stretches from the raw resources you work with to the money you receive for the products produced in each business enterprise. This chain has three links to which human creativity is applied: *resource conversion*, *product conversion*, and *marketing (or money conversion)*.

In an agricultural business, you are using human creativity and money to convert sunlight energy through plants into a saleable or consumable form. In the *product conversion* link, you turn the plants grown in the first link into a marketable form, commonly crops, livestock, or wildlife. In the *marketing (or money conversion)* link the products of the second link are marketed, and money is finally derived from the sunlight captured in the first link.

When strengthening the chain of production requires money, and profit is a goal, the proposed investments should pass the weak link check (by addressing the identified weakest link). Increasing investment in advertising (marketing link) will not profit a business that turns out a poor product (product conversion link) as much as improving that product would do. Only investment in the weak link will result in more profit at the end.

Marginal Reaction

Which action provides the greatest return toward the goal for each additional unit of money or time invested?

In the marginal reaction check, you are comparing two or more actions that have passed the other checks. You are trying to determine which of the actions will result in each additional dollar or hour of labor being invested where it provides the highest return in terms of your goals at any given moment. In other words, which action gets the "biggest bang for the buck?" No two actions can possibly give you the same return for each unit of effort (money or time) invested at that moment. So, when resources are limited, you want to select the one from which you gain the most. In doing so you will end up spending less time or money, and achieve what you want more quickly.

Gross Profit Analysis

Which enterprises contribute the most to covering the overheads of the business?

In the gross profit analysis check, you are comparing two or more enterprises by looking at the income likely to be derived from each enterprise and deducting the additional money you will have to spend to bring in that income— *the difference between money in and money out is the gross profit.* The additional money to be spent is that money you would not spend *unless* you undertook the enterprise. Through the analysis and comparison of many possible enterprises with this check, you are selecting the best enterprise, or combination of enterprises, to create profit and minimize risk.

Energy/Money, Source, and Use

Is the energy or money to be used in this action derived from the most appropriate source in terms of our holistic context?

Will the way in which the energy or money is used be in line with our holistic context?

While the gross profit analysis check helps you select enterprises based on profitability, there are other factors to consider. Not only the enterprise itself but also the secondary inputs that support it must pass the other checks that apply. This check examines both the sources and the patterns of use of the energy and money used in production. We lump money and energy together because any action contemplated usually requires one or the other, and often both.

This check helps you avoid

- actions likely to lead you into an increasing dependency on, or addiction to, fossil fuels or any other inputs; and

- actions involving an addictive use of borrowed money on which you are paying compound interest.

Sources of Energy

In terms of availability, energy sources fall into two categories: those that are abundant or unlimited, and those that are limited in supply. Solar energy that is used to grow the farmer's crops is unlimited and virtually inexhaustible; but the diesel used to fuel the farm machinery is not. We will eventually run out of secure and affordable supplies of all fossil fuels.

In terms of their effects on the environment, energy sources fall into three categories: benign, damaging, or potentially damaging. When you choose to walk or ride a bike instead of driving your car, you are choosing a more benign source of energy rather than a potentially more damaging one. So is a farmer who uses livestock to break down corn stubble in a field rather than using some type of machinery. Keeping in mind your holistic context, determine whether the source of energy you plan to use in taking an action is appropriate in your present situation.

Sources of Money

Money used to implement any action can be derived from either internal or external sources. Internal money comes from your earnings, what your business or land generates. Any time you can rely on an internal source, you are likely to be better off, but there may be many occasions when money will have to come from outside the business in order for you to move forward. When the money to be invested is derived from an external source, you need to be wary of any strings attached to it. Money from a bank or other lending institution has interest associated with it, and in most cases that interest is compound interest. If the source is a government cost-share program, it may involve management requirements that are not in line with your holistic context. It may also become addictive, and its sudden withdrawal could spell financial ruin.

Energy and Money Patterns of Use

In the second part of this check you are looking at how the energy and money will be used and whether it is appropriate in terms of your holistic context. Energy and money that are used to *build infrastructure* (e.g., buildings, machinery, fencing, transport, trained staff, a specific knowledge base, etc.), or used *cyclically*, where your money grows and energy renews with no further inputs, are preferable to *consumptive* uses that have no lasting effect. But many of your running costs, such as fuel for vehicles, or fees to your accountant, are consumptive uses, and if the action passes most of the other checks it tends to pass this one.

What you want to avoid are addictive uses of money or energy. An addictive use is one that obliges you to take the same action again and again, possibly with increasing frequency or increasing cost. Examples include the use of chemical fertilizers or the overuse of credit cards carrying high and compounding interest.

Sustainability

If we take this action, will it lead toward or away from the future resource base described in our holistic context?

This is one of the few checks that asks you to focus on a specific aspect of your holistic context. The future resource base describes the environment and behaviors that will be essential for sustaining the quality of life you desire for yourself and your descendants. Thus the sustainability check assures that actions you take to meet short-term needs also provide lasting gain—that they are socially, environmentally, and economically sound in terms of the future, as well as the present.

Your Behavior

No matter what type of business you are in, you need to consider how the perceptions of the people included in your future resource base (clients/customers and suppliers, extended family, advisers, and so on) are affected by the actions you take. If you have described yourself as honest, reliable, and professional, you want to make sure that the actions you take reflect this behavior.

The Environment

If you are a land manager, the actions you take will generally deal with one or more of the tools used to manage the four ecosystem processes. In this check, you want to determine how the proposed action or tool is likely to affect those processes. Is it likely to move you toward or away from the future landscape you've described? If you are not a land manager then any actions you take should pass through this check if they involve the consumption of products or the use of technology because their production and use will tie back to the land and affect the functioning of the four ecosystem processes

Gut Feel

How do we feel about this action now? Will it lead to the quality of life we desire? Will it adversely affect the lives of others?

The gut feel check is done last because it builds on the mental picture that has formed after passing through all the other checks. Where each of the other questions asks what you *think*, this one asks how you *feel*. And that is in large part going to be based on the values reflected in the quality of life statement embedded in your holistic context.

An action may have passed all the other checks, but you may choose not to do it because it just doesn't feel right. Or, although the action didn't pass the majority of the checks, you still feel it's the right thing to do right now.

This check also asks you to consider how an action could affect the lives of those outside your immediate whole—from the society you live in, to the greater society that comprises all humans. Pleasing everybody may seem impossible, but you can go a long way by embracing the holistic principle that the health of your particular interest is not distinct from the health of the greater whole.

APPENDIX 2
FINANCIAL PLANNING FORMS

The financial planning process described in part 1 reflects the accumulated knowledge of several decades. Lessons learned in practice have enabled the Savory Institute, with the help of its network of hubs and field professionals, to refine and improve the various planning steps and will continue to do so.

To ensure that holistic managers are kept current, the Savory Institute annually updates the material included in this book in ebooks available from http://savory.global.

The financial planning forms you'll need—Worksheet, Annual Income and Expense Plan, Livestock Production Worksheet, and Control Sheet—are reproduced here (not actual size) and are also available from http://savory.global.

WORKSHEET

Date _____

Savory

Worksheet # _____

Planning Sheet Column Reference _____

	January	February	March	April	May	June	July	August	September	October	November	December	Total
Total													

ANNUAL INCOME AND EXPENSE PLAN

Year													
Plan													
- Actual													
= Difference													
Cumulative Difference to Date													
Plan													
- Actual													
= Difference													
Cumulative Difference to Date													
Plan													
- Actual													
= Difference													
Cumulative Difference to Date													
Plan													
- Actual													
= Difference													
Cumulative Difference to Date													
Plan													
- Actual													
= Difference													
Cumulative Difference to Date													
Plan													
- Actual													
= Difference													
Cumulative Difference to Date													
Plan													
- Actual													
= Difference													
Cumulative Difference to Date													
Plan													
- Actual													
= Difference													
Cumulative Difference to Date													
Plan													
- Actual													
= Difference													
Cumulative Difference to Date													
Plan													
- Actual													
= Difference													
Cumulative Difference to Date													
Plan													
- Actual													
= Difference													
Cumulative Difference to Date													
Plan													
- Actual													
= Difference													
Cumulative Difference to Date													
PLAN TOTALS													
ACTUAL TOTALS													

LIVESTOCK PRODUCTION WORKSHEET

Savory

Property _____

Enterprise _____

Remarks: _____

Date of Plan _____

YEARS OR MONTHS

A	B	C	D	E	F	G	H		I	J	K	L	M	N
CLASS OF STOCK (Cows, Heifers, Calves, Bulls, etc.)	BIRTH % EST.	OPEN NO.	AGE	BIRTHS	MONTH	BUY	MONTH	CLASS TRANSFERS IN / OUT	DEATH	%	SALE	MONTH	CLOSE & OPEN #	AGE
1.								1						
2.								2						
3.								3						
4.								4						
5.								5						
6.								6						
7.								7						
8.								8						
9.								9						
10.								10						
11.								11						
12.								12						
13. TOTAL HEAD														

NOTE: RECORD YOUR ESTIMATES OF PERCENTAGE ACTUAL BIRTHS FOR THE VARIOUS AGE CLASSES OF BRED FEMALES IN THEIR ROWS IN COLUMN (B) ABOVE.

ANALYSIS OF PLANNED SALES AND PURCHASES

14. CLASS OF STOCK			
15. NUMBER SOLD / MONTH OF SALE			
16.			
17. AVERAGE LIVE WEIGHT			
18. MEAT PRICE PER LB (KG).			
19. INCOME PER ANIMAL			
20. WOOL/HAIR WEIGHT / MONTH OF SALE			
21. WOOL/HAIR PRICE / MONTH OF SALE			
22. WOOL/HAIR INCOME			
23. PLANNED GROSS INCOME			
24. NUMBER PLANNED TO BUY			
25. ESTIMATED PRICE/ANIMAL			
26. TOTAL COST AND MONTH			

© SAVORY INSTITUTE 2019

 Savory

CONTROL SHEET

Name _____   Date _____   Sheet # _____

Plan Column #	Amount Adverse to Plan	Amount Adverse YTD	Cause of Deviation from Plan	Proposed Action to Return to Plan	ACT

APPENDIX 3
GRAZING TWO OR MORE HERDS IN THE SAME GRAZING UNIT

Do not attempt the sophisticated practices included in appendix 3 until you have a good understanding of the planned grazing basics and confidence in your ability to implement your plans.

Although the land in a grazing unit benefits most when livestock run in a single herd, some situations may call for running two or more herds in the same grazing unit. You can do this in three ways:

1. Move separate herds among all paddocks while keeping recovery times adequate.

2. Allocate certain paddocks to each herd and plan each division as a subunit.

3. Have one herd enter a paddock as another leaves (follow-through grazing).

None of these strategies should be attempted if you have few or very uneven paddocks because grazing periods will prove to be unacceptably long.

The first option might be used when you wish to give a smaller herd, such as steers you plan to sell, a larger area to select from. The larger herd will go through those same paddocks at some point, but at higher density, which means each animal has less area to select from.

You might choose the second option because you have to separate some animals to run in a few paddocks close to the house or barn, but you want the main herd moving through the rest of the paddocks.

Although follow-through grazing is particularly tricky to plan, it sometimes fills certain needs best. For instance, when herds require different levels of nutrition, such as first-calf heifers and mature cows or, commonly in dairy operations, where you have high-performing cows, followed by average cows, then nonmilkers.

Mastering the Basics

The section that follows, on creating your plan, tells how to compute average grazing periods for each herd. Figure A3-1 presents two follow-through grazing cases. Equal paddocks cause no problem, though plants are exposed to animals for twice the grazing period of one herd. Putting two herds in Grazing Unit II could involve some heavy mathematics.

In practice the following guidelines plus careful and continual monitoring will usually serve:

- If one herd is small, you might key all moves to the larger herd. Holding a small number longer than desirable in smaller paddocks will not hurt much. (But still calculate the ADA or ADH.)

- If the moves progress from larger to smaller paddocks, the following herd can skip an occasional paddock and catch up.

- If moves progress from smaller to larger paddocks, the lead herd can skip ahead, while the following herd actually makes the first grazing of the intervening smaller paddocks.

Note that when one herd skips a paddock, you shorten recovery times. So try to plan similar paddocks in sequence.

Creating Your Plan

Before you decide to run more than one herd in a grazing unit use the method described in box 2-9 (in part 2) for deciding the number of herds. Think of the consequences to the land, livestock, and other land uses, and the degree of management each alternative requires. Then make your decision.

Record your decision on the grazing chart in row 25. For two herds of 100 and 500 animals, respectively, enter 2/100–500. If you plan a follow-through grazing, note which herd goes first (the size of the first herd will affect the performance of the following herd). If you have 50 replacement heifers followed by 250 cows—enter 2/50h–250c. For dairy animals you might have three herds—high-yield milkers, the main group, and nonmilkers—enter 3/20h–40m–34n. Use whatever abbreviations are clearest to you.

Follow-Through Grazing Planning

1	2	3	4	5	6	7
180 acres	180 acres	180 acres	180 acres	180 acres	180 acres	180 acres
12 Days	12 Days	12 Days	12 Days	12 Days	12 Days	12 Days

1	2	3	4		
140 acres	140 acres 15 Days	140 acres 15 Days	280 acres 30 Days		
15 Days	7 140 acres 15 Days	6 140 acres 15 Days	5 280 acres 30 Days		

Grazing Unit I: Follow-through grazing with 12 days in equal paddocks gives 60 days recovery.

Grazing Unit II: The follow-through plan below gives 60 – 105 days recovery in unequal paddocks.

A grazing plan for 60-day recovery times in Grazing Unit II might look like this. Note paddocks 3 and 6 are only grazed once, and 4 and 5 are grazed too long.

Herd 1 ▬▬
Herd 2 ▬▬

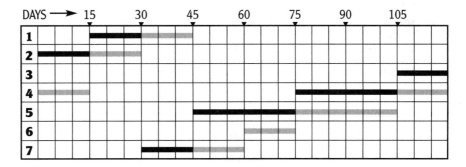

Figure A3-1. In Grazing Unit I, equal paddocks cause no problem, though plants are exposed to animals for twice as long as they would be with one herd. In Grazing Unit II, planning for two herds would likely involve a lot of arithmetic.

Calculating Average Grazing Periods

In Step 10 (Growing Season Plan) and Step 9 (Nongrowing Season Plan) of the aide memoire (found in part 2), use the following formulae to calculate average grazing periods.

Two or More Herds Using Any Paddock in the Grazing Unit

- **Growing Season Plan**

$$\frac{\text{Minimum Recovery Period}}{(\#\text{paddocks} \div \#\text{herds}) - 1} = \text{Average Minimum Grazing Period}$$

$$\frac{\text{Maximum Recovery Period}}{(\#\text{paddocks} \div \#\text{herds}) - 1} = \text{Average Maximum Grazing Period}$$

- **Nongrowing Season Plan**

    Average Grazing Period = # nongrowing days or # drought reserve days ÷ (# paddocks ÷ # herds) ÷ # selections

Example: If you had a nongrowing season of 180 days, 50 paddocks, two herds, and two selections, your average grazing period would be 3.6 days. You should include the full decimal figure for now, but you will round it off to the nearest whole number when you calculate actual grazing periods for the nongrowing period and drought reserve. Record these average days in row 28 under each month (e.g., 2.9–11.7, Growing Season plan; or 3.6, Nongrowing Season plan).

Two or More Herds with Certain Paddocks Allocated to Each Herd

- **Growing Season Plan:** In this case, you have to calculate four average grazing periods—two per herd.

    - *Herd one*

        Minimum recovery period ÷ (# paddocks allocated – 1) = Average minimum grazing period (AMGP)

        Maximum recovery period ÷ (# paddocks allocated – 1) = Average maximum grazing period (AMxGP)

- *Herd two:* Repeat the above calculations using the number of paddocks you have allocated to the second herd.

- Record all four of these average days in row 28 (e.g., 3.6–11.7/2.5–6.6). Note: If you are using this policy, go to column 3 and color-code which paddocks are to be used by each herd. This information will be needed in calculating actual grazing periods.

- **Nongrowing Season Plan:** In this case, you calculate two average grazing periods—one per herd for both the nongrowing months and the drought reserve period.

  - *Herd one*

    Average Grazing Period = # nongrowing days or # drought reserve days ÷ # paddocks allocated ÷ # selections

  - *Herd two:* Repeat the above calculations using the number of paddocks you have allocated to the second herd. Example: If you had a nongrowing season of 180 days, 100 paddocks, and two selections, and you allocated 25 paddocks to heifers and 75 paddocks to the cow herd, the average grazing period for the heifers would be 3.6 days and for the cow herd 1.2 days. You should include the full decimal figure for now, but will round it off to the nearest whole number after calculating actual grazing periods.

  - Record both figures in row 28 under each month in both the nongrowing period and drought reserve (e.g., h/3.6–c/1.2). Note: If you are using this policy, go to column 3 and color code which paddocks are to be used by each herd. (This information will be needed in Step 10, Growing Season Plan, or Step 9, Nongrowing Season Plan).

Two Or More Herds on Follow-Through Grazing

- **Growing Season Plan:** If you are using *one recovery period* calculate one grazing period. That grazing period will be used by each herd.

For example, if you had two herds and a two-day grazing period, the first herd would be in each paddock two days on average, and the second herd the next two days. The paddock would thus have animals in it for four days (two herds × two days).

Recovery period ÷ (# paddocks – # herds) = Average Grazing Period

If you are using *a range of recovery periods* calculate two average grazing periods. The same now applies, except the lower grazing period figure will guide the speed of moves in fast growth and the longer figure in poor growing conditions. This is very difficult to manage if you lack experience or if paddocks are uneven in size or quality.

$$\frac{\text{Minimum Recovery Period}}{(\text{\# paddocks} - \text{\# herds})} = \text{Average Minimum Grazing Period}$$
(for each herd)

$$\frac{\text{Maximum Recovery Period}}{(\text{\# paddocks} - \text{\# herds})} = \text{Average Maximum Grazing Period}$$
(for each herd)

Record these average grazing period(s) in row 28 (e.g., 3.5, or 1.2–4.3).

- **Nongrowing Season Plan:** You will only calculate one grazing period, but that grazing period will be used by each herd. For example, if you had two herds and a two-day grazing period, the first herd would be in each paddock two days on average, and the second herd the next two days. The paddock would thus have animals in it for four days (two herds × two days) on average. (One selection is the norm with two or more herds on follow-through grazing.)

  # nongrowing days, or # drought reserve days ÷ # paddocks = Average Grazing Period (for each herd)

  Example: If you had a nongrowing season of 180 days, 100 paddocks, and two herds on follow-through, your grazing period would be 1.8 days for each herd. Each herd would spend 1.8 days on average in each paddock. You should include the full decimal figure for now, but you will round it off to the nearest whole number after calculating actual grazing periods. Record both figures in row 28 under each month for the nongrowing period and drought reserve (e.g., 1.8–1.8).

Calculating Actual Grazing Periods

Factor in grazing unit productivity as you did for a single herd in Step 8 (Growing Season Plan) and Step 7 (Nongrowing Season Plan) of the aide memoire, and use the same formula in each case:

Paddock rating × Average grazing period ÷ Average paddock rating = Actual Grazing Period

Distinguishing One Herd from Another

When plotting the grazings in step 12 (Growing Season Plan) or step 11 (Nongrowing Season Plan), use symbols or special lines to distinguish one herd from another (as shown in figure A3-1) and note which symbol/line represents each herd at the bottom of the chart in the "Remarks" section.

APPENDIX 4
A MORE REFINED METHOD FOR CALCULATING GRAZING PERIOD ADJUSTMENTS BASED ON HERD SIZE CHANGES

Do not attempt the sophisticated practices included in appendix 4 until you have a good understanding of the planned grazing basics and confidence in your ability to implement your plans.

When you are creating a nongrowing season plan and herd size varies dramatically you might consider using a more refined method for adjusting the grazing periods than suggested in step 10 of the nongrowing season aide memoire (found in part 2). This method involves a number of calculations. To start, you determine a herd size adjustment factor, which is based on the difference between average and actual herd sizes in both the non-growth and the drought reserve periods. Then you adjust the grazing periods plotted in step 11 based on the herd size adjustment factor.

1. Calculate the *average* herd size for the non-growing months and then for the drought reserve period.

   • Multiply the number of standard animal units (SAUs) in the herd (row 34) by the number of days the herd remains the same size (SAUs × days of same herd size). Record the resulting animal-day (AD) figure in brackets in row 34. When herd size changes, do the same thing again, and continue doing so until the end of the planned period (nongrowth or drought reserve).

   • Add up all the AD figures and divide by the total days in the planned period. This gives you the average herd size. Record this figure in the "Remarks" section.

   Example—Nongrowing Months: Assume the planned period is 200 days. Herd size during the first 60 days is projected to be 450 SAUs (450 × 60 = 27,000 ADs), then calves are weaned, dry cows are culled, and herd size is reduced to 300 SAUs for 45 days (13,500 ADs), then year-

ling heifers are added to the herd, and size goes back up to 375 SAU for 95 days (35,625 ADs). Total ADs accumulated is therefore 76,125. That figure divided by 200 (the total days in the planned period) gives an average herd size of 381 SAU:

$$(27,000 + 13,500 + 35,625) \div (60 + 45 + 95) = (76,125 \div 200) = 381 \text{ SAU}$$
Average Herd Size

Example—Drought Reserve Period: Assume the drought reserve is 60 days. Herd size during the first 15 days is projected to be 375 SAU (5,625 ADs), but will increase to 450 SAU during the next 30 days after 75 pregnant cows are bought (13,500 ADs). It will increase to 500 SAU over the last 15 days as calves start to come (7,500 ADs), for a total of 26,625 ADs. That figure divided by 60 (the total days in the drought reserve period) gives an average herd size of 444 SAU.

2. Calculate an *adjustment factor* for each different herd size by dividing the *average* herd size by the *actual* herd size:

   Average herd size ÷ Actual herd size
   = Adjustment factor

   Example—Nongrowing Months: If the average herd size is 381 SAUs, then in the first 60-day period, when actual herd size is 450 SAUs, the adjustment factor would be 0.9 (381 ÷ 450). In the next 45-day period when herd size decreases to 300 SAUs, the factor becomes 1.3 (381 ÷ 300). In the final 95-day period, when herd size increases to 375 SAUs, no adjustment is required (381 ÷ 375 = 1).

   Example—Drought Reserve Period: If the average herd size is 444 SAUs, then in the first 15-day period when actual herd size is 375 SAUs, the adjustment factor would be 1.2 (444 ÷ 375). In the next 30-day period when herd size increases to 450 SAUs, no adjustment is required (444 ÷ 450 = 1). In the last 15 days when herd size increases to 500 SAUs, the factor is 0.9 (444 ÷ 450). Obviously, the greater the fluctuation in herd numbers, the greater the range of adjustment factors.

3. Record the herd size adjustment factors (in brackets) to the right of the herd size figures in row 25. Then multiply each of these figures by the actual grazing period in column 4 (or the drought reserve column to its right). Round the total off to the nearest whole day (or half day) and pencil in the adjusted grazing period line. Occasionally, herd size will change during a planned grazing period in an individual paddock. This is especially common when you have few paddocks and therefore long grazing periods. The following procedure should be followed when that happens, especially if herd size is going to change dramatically and if the calculated actual grazing period is very long.

4. Calculate how many ADs would be harvested from the paddock based on the average herd size. For example, assume the actual grazing period recorded in column 4 is 25 days, and the average herd size for the planned period is 300. That would result in 7,500 ADs harvested with the average herd size. Make a note of that number so you don't forget it.

5. Next, calculate how many ADs will be accumulated with the actual herd size up to the time the herd size changes. For example, if herd size is 200 SAUs in the first 10 days (of a 25-day grazing period), total ADs harvested would be 2,000. That leaves 5,500 ADs in that paddock for the remainder of the grazing period (7,500 − 2,000). If herd size after day 10 is planned to increase to 280 SAUs, then there are 20 days of grazing left in the paddock (5,500 ADs ÷ 280 SAUs). That means the grazing period can be extended from 25 to 30 days (10 + 20), which you should pencil in on the chart.

APPENDIX 5
GRAZING PLANNING FORMS

The aide memoires described in part 2 are updated annually to reflect the lessons learned by practitioners worldwide in the ebooks published by the Savory Institute. Some of these updates also dictate revisions in the Grazing Plan and Control Chart that accompany the aide memoires. The Savory Institute produces the chart in several formats: one large-print version that can be copied on two sheets, one with twenty paddocks, and one with thirty-five paddocks. Check periodically with your local Savory Network Hub or with http://savory.global to ensure that you have the most up-to-date aide memoires and charts.

The twenty-paddock single-page version of the Grazing Plan and Control Chart reproduced here (in reduced form) can be ordered, along with the other versions listed above from http://savory.global. The worksheet shown in appendix 2 is also used in grazing planning and can also be ordered from savory.global.

Grazing Plan & Control Chart
(Livestock/Wildlife/Crops/Other Uses)

Savory

YEAR _____ SEASON _____ GRAZING UNIT _____

COLUMNS

1	2	3		4	5	6	7	8
	Estimated Relative Paddock Quality	Paddocks		Minimum	Estimated Available		Nongrowing Season	
Actual ADA(H) (past season)		Size	Number/Name	Maximum Guidelines	ADA(H)	ADs	Planned Demand ADA(H)	Total Yield ADA(H)

21. Rainfall
22. Snow
23. Growth Rate (F/S/0)
24. Supplement or Feed—Type and Amount
25. Number/Size of Herds
26. Paddocks Available
27. Recovery Period(s) or Number Selections
28. Avg GP or AMPG/AMxGP

Type of Animals	No.	Average Weight	% Unit	Total Units	No.	Average Weight	% Unit	Total Units	No.	Average Weight	% Unit	Total Units	No.	Average Weight	% Unit	Total Units	No.	Average Weight	% Unit	Total Units	No.	Average Weight	% Unit	Total Units
29.																								
30.																								
31.																								
32.																								
33.																								

34. Total SAUs
35. Grazing Unit Size:
36. Stocking Rate
37. Avg Annual Precipitation:
38. Season Total Precipitation

Remarks:

A. Estimated Total ADs (Livestock/Wildlife) _____
B. Estimated Days of Non Growth _____
C. Days of Bulk Feeding _____
D. Days of Drought Reserve Required _____
E. Total Days Grazing Required _____
F. Estimated Carrying Capacity _____

SUMMARY LIVESTOCK AND LAND PERFORMANCE

Calving/Lambing/Kidding _____ %
Avg Weaning Weight _____ Age _____ (Mths)
Daily Weight Gains _____ Lbs(Kgs)
(Growing)   (Nongrowing)

ADA(H)/inch (mm) Rainfall _____
Total Yield (sold) per acre(hectare) _____ Lbs(Kgs)

© **SAVORY** INSTITUTE 2020

The following sections describe the reasoning behind the interpretations given for the five monitoring scenarios in part 3 under "Monitoring Your Land."

Scenario 1

Decreasing bare ground is desirable, with broken soil increasing and mature capping down—all this suggests a fair amount of animal impact. Litter increasing is desirable and could be due to the animal impact that has increased since last year. But the fact that plant spacing has not changed while old oxidizing grass has disappeared suggests that animal impact is still low (and partial rest is still too high). The land is improving, but slower than planned for and needs action.

Scenario 2

Bare ground increasing while mature capping decreases, broken soil increases (both indicating some effectiveness of animal impact), and litter decreasing is not desirable. Erosion could well increase. Decreasing plant spacing suggests effective animal impact. No overgrazing or overresting is desirable. The overall probability here is that the stocking rate is too high, but all else seems to be in line with plans. Overstocking is indicated by disappearance of litter, as livestock have probably eaten it.

Scenario 3

Bare ground decreasing ties in with broken soil increasing while mature capping decreases and litter increases. The increase in litter will be assisted by both closer plant spacing and an increasing number of grass plants, and all seems to be in line with the planned changes.

Scenario 4

No change in bare soil, mature capping, broken soil, or plant spacing suggests inadequate animal impact or too high a degree of partial rest. That this is combined with an increase in overrested plants and new weeds or forbs suggests that the stocking rate is also too low.

Scenario 5

Decreasing bare ground, mature capping, and plant spacing all suggest that animal impact has been effective. Litter increasing also ties in with the effectiveness of the animal impact being applied. However, it is disturbing that some plants are showing visible signs of overgrazing while others are overresting and that weeds are increasing. These signs indicate that the manager either has started to rotate the livestock or, if he or she is still planning the grazing, has not been paying attention to recovery periods and daily plant growth rates.

APPENDIX 7
ECOLOGICAL MONITORING FORMS

The two ecological monitoring procedures described in part 3 have undergone many years of refinement. Practical experience will continue to lead to further innovations. To ensure that you keep current, the Savory Institute annually updates the material included in this book in ebooks available from http://savory.global.

The ecological monitoring forms reproduced here and available from savory.global include the following:

- Ecological Monitoring Data—Basic
- Ecological Monitoring Analysis—Basic
- Ecological Monitoring Data—Comprehensive
- Ecological Monitoring Summary—Comprehensive
- Ecological Monitoring Analysis—Comprehensive

ECOLOGICAL MONITORING DATA - BASIC
(5 Needed per Transect)

Property _____ **Transect/Plot #** _____ **Photo #s** _____

Date _____ **Examiner(s)** _____

1. **Soil Surface**
   Describe the nature of the bulk of the soil surface between plants. (Is it bare, capped, broken, covered with litter, covered with algae and lichen, hard, soft, porous, etc? Are there signs of soil movement/erosion, such as pedestaling, siltation in low points, etc.?)

2. **Animal Sign**
   What signs of animal life are present (small or large animals, birds, insects, reptiles)?

3. **Litter**
   If there is litter present, describe its quality/condition (fresh, old. or breaking down so it is hard to distinguish where litter ends and soil begins.)

4. **Perennial Grass Condition**
   if perennial grasses are present, describe their condition. (Are they healthy, mature, young, seedlings, dead/dying, overrested, overgrazed?)

5. **Grass Species**
   List grass species in the plot if you know their names.

6. **Other Plants**
   List or comment on other non-grass plant species present (legumes, forbs, etc.)

7. **Points of Interest**
   Note any other points of interest, including things that might not show well in the photo.

ECOLOGICAL MONITORING ANALYSIS - BASIC

(Use 1 per Transect)

Property _____ **Transect/Plot #** _____ **Photo #s** _____

Date _____ **Examiner(s)** _____

1. **What are we trying to achieve in the area surrounding this transect?**

   Community Dynamics

   Water Cycle

   Mineral Cycle

   Energy flow

2. **What progress have we made this year, compared to last year?**

   Community Dynamics

   Water Cycle

   Mineral Cycle

   Energy flow

3. **What natural or management factors might have influenced what we are seeing on the ground?**

4. *If adverse changes have occurred or no change, where change was planned:* **What is the underlying cause (what tools have been applied, and how have we applied them?)**

5. **What are we going to change in this next year to keep our land moving toward the future landscape described in our holistic context?**

© **SAVORY** INSTITUTE 2019

ECOLOGICAL MONITORING DATA - COMPREHENSIVE

Property _____ Transect _____ Photo #s _____ Date _____

Examiner(s): _____

Other Comments:

	AT DART ENTRY POINT						6-In (15-cm) CIRCLE AROUND POINT																	DESCRIBE NEAREST PERENNIAL																			
	What Dart Point Hit (Must Check One)					Yes	Soil Surface (Must Check One)					Evidence of Check if 'Yes'			What It Is (Must Check One)							Its Habitat (Check One)			Its Age (Must Check One)					Its Form (Must Check One)					Its Species (If Known)
Throw Number	Bare Soil	Litter 1	Litter 2	Rock	Plant Base	Canopy Above Point	Mature	Immature	Recent	Broken	Covered	Animal Sign: I,W,B,S,L	Annuals Present	Soil Movement	Grass: (C), (W), or (Y)	Rush or Sedge	Forb	Shrub	Tree	Distance to it (in/cm)	Dry	Middle	Wet	Seedling	Young	Mature	Decadent (Dying)	Resprout	Normal	Overrested	Overgrazed	Overbrowsed	Dead	
1.																																			
2.																																			
3.																																			
4.																																			
5.																																			
6.																																			
7.																																			
8.																																			
9.																																			
10.																																			
11.																																			
12.																																			
13.																																			
14.																																			
15.																																			
16.																																			
17.																																			
18.																																			
19.																																			
20.																																			
21.																																			
22.																																			
23.																																			
24.																																			
25.																																			
26.																																			
27.																																			
28.																																			
29.																																			
30.																																			
31.																																			
32.																																			
33.																																			
SUBTOTAL																																			

ECOLOGICAL MONITORING SUMMARY - COMPREHENSIVE

Property _____   Transect/Plot # _____   Photo # s _____   Date _____

Examiner(s) _____

Nearest Perennial	%
Cool Season Grass	
Warm Season Grass	
Year-Round Green Grass	
Rush or Sedge	
Forb	
Shrub	
Tree	
Average Distance (Inches /cm)	
Dry	
Middle	
Wet	
Seedling	
Young	
Mature	
Decadent (Dying)	
Resprout	
Normal	
Overrested	
Overgrazed	
Overbrowsed	
Dead	

(CHARACTERISTICS / AGE & FORM)

Soil Surface	%
Bare Soil	
Litter 1	
Litter 2	
Rock	
Plant Base	
Canopy	
Mature	
Immature	
Recent	
Broken	
Covered	
Insects	
Worms	
Birds	
Small Animals	
Large Animals	
Annuals	
Soil Movement	

(COVER & CAPPING / EVIDENCE)

REMARKS

Type	Plant Name or Species	No - %
GRASSES		
	Total Grasses : No. - %	
RUSH		
	Total Rushes & Sedges : No. - %	
FORBS		
	Total Forbs : No. - %	
SHRUBS		
	Total Shrubs : No. - %	
TREES		
	Total Trees : No. - %	

ECOLOGICAL MONITORING ANALYSIS - COMPREHENSIVE

(Use 1 per Transect)

Property _____ **Transect/Plot #** _____ **Photo #s** _____

Date _____ **Examiner(s)** _____

1.   What are we trying to achieve in the area surrounding this transect?

Community Dynamics
Water Cycle
Mineral Cycle
Energy flow

2.   What progress have we made this year, compared to last year?

Community Dynamics
Water Cycle
Mineral Cycle
Energy flow

3.   What natural or management factors might have influenced what we are seeing on the ground?

4.   _If adverse changes have occurred or no change, where change was planned:_ What is the underlying cause (what tools have been applied, and how have we applied them?)

5.   What are we going to change in this next year to keep our land moving toward the future landscape described in our holistic context?

GLOSSARY

A number of new words and phrases are associated with Holistic Management, particularly for those aspects related to the use of grazing animals to regenerate landscapes. This glossary defines many of those new terms along with a few others that involve new ways of looking at the land in order to judge its condition.

animal-days per acre (ADA) or hectare (ADH). A term used to express simply the volume of forage taken from an area in a specified time. It can relate to one grazing in a paddock or several, in that more grazings than one can be added to give a total ADA or ADH figure. If you're running a breeding herd, or mixed species herd, you will need to convert animal numbers to standard animal units (see definition below) to better assess the volume of forage required to feed the herd or, following a grazing, the volume of forage taken. The ADA or ADH figure is arrived at by a simple calculation as follows:

$$\frac{Animal\ numbers\ (in\ SAU) \times days\ of\ grazing}{Area\ of\ land\ acreas\ or\ hectares} = ADA\ or\ ADH.$$

Note: If you're running a single class of animals, such as summer yearlings, you can forgo the conversion to standard animal units (SAU).

animal impact. The sum total of the direct physical influences animals have on the land—trampling, digging, dunging, urinating, salivating, rubbing, etc. Most commonly achieved with herding animals in high concentration. The larger the herd, the greater the effect.

biodiversity. The diversity of plant and animal species—and of their genetic material and the age structure of their populations—within a given community.

biomass. The mass, or volume, of life—plants, animals, and microorganisms.

brittleness scale. All terrestrial environments, regardless of total rainfall, fall somewhere along a continuum from nonbrittle to very brittle. For simplicity, we refer to this continuum as a 10-point scale—1 being nonbrittle and 10 being very brittle. Completely nonbrittle environments are characterized by (1) reliable precipitation regardless of volume; (2) good distribution of humidity throughout the year as a whole; (3) a high rate of biological decay in dead plant material, which is most rapid close to the soil surface (thus dead trees rot at their bases and topple over relatively quickly); (4) speedy development of new communities on any bare surface; (5) the development of complex and stable communities even where they are not physically disturbed for many years. In such environments it is virtually impossible to produce or maintain millions of acres where the ground between plants is bare, other than on croplands that are continually exposed by machinery.

Very brittle environments, on the other hand, are characterized by (1) unreliable precipitation, regardless of volume; (2) poor distribution of humidity through the year as a whole; (3) chemical (oxidation) and physical (weathering) breakdown of dead plant material, generally slow and from the upper parts of plants downward (thus dead trees remain standing for many years); (4) very slow development of communities from bare soil surfaces unless physically disturbed; (5) soil surfaces that can be covered with algae and lichens for centuries unless adequately disturbed. In such environments it is very easy to produce millions of acres where the ground between plants is bare or algae/lichen capped, merely by resting the land excessively, burning it frequently, or by overgrazing many grass plants. Such areas tend to maintain biodiversity and stability only when they receive adequate disturbance periodically.

capping, immature. A soil surface that has sealed with the last rainfall and on which there is no visible sign yet of successional movement. Capping is initiated by raindrop action on an exposed soil. The energy from the raindrop breaks crumb structure and frees fine soil particles. These in turn seal the surface so the soil cannot respire easily. Some people use the term *crusting* instead of *capping*.

capping, mature. An exposed soil surface on which succession has proceeded to the level of an algae-, lichen-, and/or moss-dominated community and has stalled at that level. If not adequately disturbed, such communities can remain in this state for centuries provided the soil is level enough to inhibit erosion by water.

closed plan. The grazing plan created for the slow- or nongrowing months of the year, including the time reserve planned for drought. In this plan you ration out the forage over the months ahead to a theoretical end point, which should be a month or more after your most pessimistic estimate of when new growth could occur.

community dynamics. The development of communities of living organisms. This process is ongoing due to the constant interplay of species, changing composition, and changing microenvironment. However, the greater the biodiversity within a community, the more complex, and thus the more stable it tends to be.

crumb structure. A soil that has good crumb structure is made up largely of aggregates or crumbs of soil particles held together when wet or dry with glue provided by decomposing organic matter. The space around each crumb provides room for water and air, and this in turn promotes plant growth.

desertification. A process characterized by a loss of biodiversity, plant mass, and soil cover. Symptoms include increased incidence of flood and drought, declining levels of soil organic matter, increased soil surface exposure, and erosion.

drought (or time) reserve. The number of days or months of grazing you plan to reserve in case of drought. This time period would extend from the end of an average nongrowing season to a month or so past the date you expect to receive new growth. Base your estimate on the longest possible weather records available.

effective recovery period. One in which a severely grazed plant has actually grown new leaves and stems and restored energy reserves in, depending on the species, reestablished roots, stem bases, or crowns. This can only occur under active growing conditions.

forbs. Tap-rooted herbaceous plants, often referred to as weeds.

graze/trample-to-recovery ratio. The number of days animals are on a piece of land, divided into the number of days they will be off it before returning. Generally, the shorter the grazing periods and the longer the recovery periods—or the higher the ratio—the better the performance of both land and animals, although there is more to it than that. Bear these general points in mind in deciding: A very high ratio, especially during the growing season, when all recovery time is effective in terms of growth, will tend to result in improved soil condition due to considerably more root growth and organic matter buildup. However, when soils are leached, rainfall is high, or excessive lignin develops in plants, individual livestock performance could drop. This drop is somewhat offset by the very short grazing periods necessary to achieve the high ratio. Generally, a high ratio will allow for a higher stocking rate, and thus higher financial return per acre or hectare. A doubling of stocking rate generally leads to higher profitability unless individual performance drops forty percent or more, which normally does not happen.

grazing, frequent. Grazing that takes place with short intervals between the actual grazings on the plant. With most plants, frequent grazing is not harmful as long as the defoliation is light.

grazing selections. The number of times you plan to have animals move through a paddock in the *nongrowing season* (when there is no significant regrowth between each grazing).

grazing, severe. Grazing that removes a high proportion of the plant's leaf in either the growing or the nongrowing season. In the growing season this causes a temporary setback in the plant's growth. In the more brittle environments severe grazing at some time during the year is generally beneficial to most bunched perennial grass plants, and especially those with growth points, or buds, at their bases. Most herding animals are severe grazers including cattle, sheep, and goats.

grazing unit. An area of subdivided land that is *planned as one unit* to regulate the time that plants and soils are exposed and reexposed to grazing and trampling. A grazing unit is always planned on one Grazing Plan and Control Chart, and generally includes only one herd. A grazing unit can incorporate any number of subdivisions (paddocks) that will receive varying recovery periods—all of which must be visible on one chart. Several smaller grazing units can be combined to form one large grazing unit for planning purposes.

herd effect. The impact on soils and vegetation produced by a large herd of animals in high concentration or in an excited state. Herd effect is not to be confused with stock density, as they are different, although often linked. You can have high herd effect with very low stock density (e.g., the bison of old that ran in very large herds at very low stock density, as the whole of North America was the paddock). You can have high stock density with no herd effect, such as when two or three animals are placed in a one-acre paddock. At ultra-high densities, the behavior of livestock will often change adequately to provide herd effect (see *ultra-high-density grazing*).

Note: Herd effect is the result of a change in animal behavior and usually has to be brought about by some actual management action—stimulating the behavior change with the use of an attractant, or crowding animals to ultra-high density. Herd effect generally provides the (animal) impact needed to break up a capped soil surface, to compact soil enough to get good seed-to-soil contact, and to trample dead plant material to the ground where it provides soil cover and slows water movement and erosion. Applied too long or too frequently it tends to pulverize most soils and cause excessive compaction.

Holistic Planned Grazing. The planning of livestock moves using a step-by-step guide that caters simultaneously for many variables: animal behavior, breeding, performance, wildlife needs, other land uses, weather, plant growth rates, poisonous plants, dormant periods, droughts, and so forth. The purpose of such planning is to utilize livestock to bring about a future landscape that is in line with the holistic context created by the managers.

low-density grazing (sometimes referred to as patch or selective grazing). This refers to the grazing of certain areas while others nearby are left ungrazed and on which plants become old, stale, and moribund. Normally it is caused by stock grazing at too low a stock density, too small a herd, or a combination of these, with too short a time in a paddock. Once it has started, even by only one grazing, it tends to get progressively worse, as the nutritional contrast between regrowth on grazed areas and old material on ungrazed areas increases with time. The common remedy calls for holding stock longer to force them to graze everything equally (nonselectively), but this is a bad mistake and results in stock stress and poor performance. It is low-density grazing and is corrected by increasing density and generally shortening time to avoid stock stress.

mineral cycle. The cycling of mineral nutrients from soil to aboveground plants and animals and back to the soil again. A healthy and productive environment will promote the movement of minerals from deep soil layers to aboveground plants with a minimum of mineral loss from soil erosion or mineral leaching.

open-ended plan. The grazing plan created for the growing months of the year. In this plan you are trying to grow as much forage as possible, and you do not have to plan to a specific date. The plan remains open because you don't know when growth will end or exactly how much forage will grow before that date.

overgrazing. When a perennial grass plant bitten severely *in the growing season* gets bitten severely again while using energy it has taken from its crown, stem bases, or roots to reestablish leaf. Generally, this results in the eventual death of the plant. In intermediate stages it results in reduced production from the plant. Overgrazing commonly occurs at three different times: (1) when the plant is exposed to the animals for too many days and they are around to regraze it as it tries to regrow, (2) when animals move away but return too soon and graze the plant again while it is still using stored energy to reform leaf, or (3) immediately following dormancy when the plant is growing new leaf from stored energy.

overrested plant. A bunched perennial grass plant that has been rested so long that accumulating dead material prevents light from reaching growth points at the plant's base, hampering new growth and eventually killing the plant. Overrest occurs mainly in brittle environments where, in the absence of large herbivores, most old material breaks down gradually through oxidation and weathering rather than rapidly through biological decay.

paddock. A division of land within a grazing unit, either fenced off, or demarcated for herding. Several or many paddocks together make up a grazing unit, provided they are planned as one unit on a grazing chart. The American term *pasture*, when used to define an area of land, is synonymous with *paddock*.

pasture. A planted grass or other forage crop. In the United States, *pasture* can refer to grass on the range, a planted grass sward, or a division of land. To avoid confusion, we use the word *paddock* to refer to a division of land, *pasture* to refer to a planted grass sward or forage crop.

rest, partial. Takes place when grazing animals are on the land but without a full complement of pack-hunting predators to maintain bunching-herd behavior most of the time. As a result, animals barely disturb soil surfaces and trample very little material onto the ground. It commonly results in damaged algae/lichen communities but no successional advance to more complex, stable communities.

rest, total. Prolonged nondisturbance to soils and plant/animal communities. A lack of physical disturbance and/or fire.

rest-tolerant grasses. Perennial grasses able to thrive under rest in very brittle environments. Commonly, such plants have some growth points, or buds, well above ground along their stems where unfiltered sunlight can reach them; or they are short in stature or sparsely leafed, enabling unfiltered light to reach their ground-level growth points. In the past, such grasses tended to be found in steep gorges and other sites large grazing animals did not frequent. Today, however, because overgrazing is believed to be linked to animal numbers, and numbers have been reduced, grasses that don't require grazing to thrive have an advantage. Rest-tolerant grasses can be found all over.

rotational grazing. Grazing in which animals are rotated through a series of paddocks, generally on some flexible basis, but without any planning that caters for the many variables inherent in the situation.

standard animal unit (SAU). A single class of animal against which you rate all the other classes/species of animals in a herd to better assess the herd's forage requirements. A mature cow is often used as the standard, where a cow equals 1 animal unit. A cow and her calf would be 1.5 units, a bull 2 units; 5 sheep would equal 1 animal unit, etc. SAU × days = animal-days (the amount of forage the standard animal would eat in a day).

stock density. The number of animals run on a subunit (paddock) of land at a given moment in time. This could be from a few minutes to several days. Usually expressed as the number of animals (of any size or age) run on one acre or hectare.

stocking rate. The number of animals run on a unit of land, usually expressed in the number of acres or hectares required to run one full-grown animal throughout the year or part thereof.

strip grazing. The grazing of animals on narrow strips of land generally behind a frequently moved electric fence. In some cases, different areas are strip grazed within a paddock.

succession. An important aspect of community dynamics, succession describes the stages through which biological communities develop. As simple communities become ever more diverse and complex, succession is said to be advancing. When complex communities are reduced to greater simplicity and less diversity, succession is set back. If the factors that set it back are removed, succession will advance once again.

transect. A selected piece of land on which data are gathered and/or photos taken year after year to monitor any changes arising from management practices.

ultra-high-density grazing. The grazing of livestock in such a manner that they are at extremely high stock densities throughout the day. Generally, these densities are achieved either by herding the stock, by enclosing them in a small area with the use of moveable fencing, or in utilizing a combination of both. The aim is to induce herd effect, and thus high animal impact, over most of the land most of the time.

water cycle. The movement of water from the atmosphere to the soil (or the oceans) and eventually back to the atmosphere again. An effective water cycle is one in which plants make maximum use of rainfall. Little evaporates directly off the soil, and any runoff causes no erosion and remains clear. A good air-to-water balance exists in the soil, enabling plant roots to absorb water readily. For the water cycle to be effective in brittle environments, the soil must be covered with living plants and/or litter, as vast amounts of water are lost through the bare, exposed soil between plants.

ABOUT THE AUTHORS

Jody Butterfield is a former journalist specializing in agriculture and the environment. She is a cofounder of the Savory Institute.

Sam Bingham is a Savory Network Accredited Educator. He is the author of five books and numerous articles on the land and the people who make a living from it.

Allan Savory is a former wildlife biologist and farmer who conceived of and developed Holistic Management. He is a cofounder and president of the Savory Institute.

ABOUT SAVORY GLOBAL

Savory Global facilitates large-scale regeneration of the world's grasslands through Holistic Management. Savory's goal is to regenerate one billion hectares of grasslands through its global network to mitigate climate change and enhance water and food security. Savory Global includes the Savory Institute, the Savory Global Network, and the Land to Market program.

The **Savory Institute** equips the global network with innovative tools and curricula, informs policy, engages commercial markets, increases public awareness, and conducts relevant research on ecological, social, and financial outcomes. Additionally, the Institute develops and incubates these programs through its experiential learning site, West Bijou Ranch, in Colorado, USA.

The **Savory Global Network** comprises Savory Hubs, Accredited Professionals, and Regenerating Members. It advocates, trains, and supports land managers to regenerate land in their own context through Holistic Management.

The Land to Market program, deployed by the global network is the world's first outcome-based verified **regenerative sourcing** solution. The program connects conscientious brands, retailers, and consumers directly to producers who are regenerating their land. The more farmers and ranchers the hubs train, support, and verify, the more grasslands regenerated, and the greater the environmental, economic, and social impact globally.